Chapman & Hall/CRC
Computer Science and Data Analysis Series

CHAIN EVENT GRAPHS

Rodrigo A. Collazo

Naval Systems Analysis Centre
Rio de Janeiro, Brazil

Christiane Görgen

Max Planck Institute for Mathematics in the Sciences
Leipzig, Germany

Jim Q. Smith

University of Warwick
United Kingdom

CRC Press
Taylor & Francis Group
Boca Raton London New York

CRC Press is an imprint of the
Taylor & Francis Group, an **informa** business
A CHAPMAN & HALL BOOK

Chapman & Hall/CRC
Computer Science and Data Analysis Series

The interface between the computer and statistical sciences is increasing, as each discipline seeks to harness the power and resources of the other. This series aims to foster the integration between the computer sciences and statistical, numerical, and probabilistic methods by publishing a broad range of reference works, textbooks, and handbooks.

Proposals for the series should be sent directly to one of the series editors above, or submitted to:

Chapman & Hall/CRC
Taylor and Francis Group
3 Park Square, Milton Park
Abingdon, OX14 4RN, UK

Published Titles

Semisupervised Learning for Computational Linguistics
Steven Abney

Visualization and Verbalization of Data
Jörg Blasius and Michael Greenacre

Chain Event Graphs
Rodrigo A. Collazo, Christiane Görgen, and Jim Q. Smith

Design and Modeling for Computer Experiments
Kai-Tai Fang, Runze Li, and Agus Sudjianto

Microarray Image Analysis: An Algorithmic Approach
Karl Fraser, Zidong Wang, and Xiaohui Liu

R Programming for Bioinformatics
Robert Gentleman

Exploratory Multivariate Analysis by Example Using R
François Husson, Sébastien Lê, and Jérôme Pagès

Published Titles cont.

Bayesian Artificial Intelligence, Second Edition
Kevin B. Korb and Ann E. Nicholson

Computational Statistics Handbook with MATLAB®, Third Edition
Wendy L. Martinez and Angel R. Martinez

Exploratory Data Analysis with MATLAB®, Third Edition
Wendy L. Martinez, Angel R. Martinez, and Jeffrey L. Solka

Statistics in MATLAB®: A Primer
Wendy L. Martinez and MoonJung Cho

Clustering for Data Mining: A Data Recovery Approach, Second Edition
Boris Mirkin

Introduction to Machine Learning and Bioinformatics
Sushmita Mitra, Sujay Datta, Theodore Perkins, and George Michailidis

Introduction to Data Technologies
Paul Murrell

R Graphics
Paul Murrell

Correspondence Analysis and Data Coding with Java and R
Fionn Murtagh

Data Science Foundations: Geometry and Topology of Complex Hierarchic Systems and Big Data Analytics
Fionn Murtagh

Pattern Recognition Algorithms for Data Mining
Sankar K. Pal and Pabitra Mitra

Statistical Computing with R
Maria L. Rizzo

Statistical Learning and Data Science
Mireille Gettler Summa, Léon Bottou, Bernard Goldfarb, Fionn Murtagh, Catherine Pardoux, and Myriam Touati

Music Data Analysis: Foundations and Applications
Claus Weihs, Dietmar Jannach, Igor Vatolkin, and Günter Rudolph

Foundations of Statistical Algorithms: With References to R Packages
Claus Weihs, Olaf Mersmann, and Uwe Ligges

CRC Press
Taylor & Francis Group
6000 Broken Sound Parkway NW, Suite 300
Boca Raton, FL 33487-2742

First issued in paperback 2020

Version Date: 20171221

ISBN-13: 978-0-367-57231-0 (pbk)
ISBN-13: 978-1-4987-2960-4 (hbk)

Visit the Taylor & Francis Web site at
http://www.taylorandfrancis.com

and the CRC Press Web site at
http://www.crcpress.com

To our families

Contents

Preface

Jim Smith became aware of the inferential power of statistical models based on graphs after a series of talks given at the University of London by the great practical Bayesian Lawrence Phillips. Soon after that time an enormous momentum from academics built up on the use of graphical models. In particular, the *Bayesian network* was developed to give exciting new ways of representing relationships in real problems informed by data and expert judgements. This provided excellent ways of explaining relationships embedded within observational series. It was only in about 2000, after being an invited discussant to a paper and then talking to Glenn Shafer, that it dawned on Jim Smith that graphical representations of *event trees* might actually be an even more powerful tool than the Bayesian network for representing and studying discrete problems. Jointly with his two former PhD students Rodrigo Collazo and Christiane Görgen, he has now written this book to demonstrate that this is indeed the case.

We three authors hope to demonstrate over the next two hundred pages how interesting and useful this new class of discrete models is and why we believe that many researchers and practitioners would benefit from using the techniques we have developed. Many of our methods can now be experimented with using the freely available `R`-package `ceg` [18].

This book requires some basic knowledge of probability theory and a little university training in mathematics. We have tried to keep the text as non-technical as we can so that it is accessible to as wide an audience as possible. The more complex work lying behind this development has been published in a series of peer-reviewed papers which are cited throughout the text. Many of the original ideas are here illustrated through basic examples. Therefore we believe that the material should be easily accessible to students in their third year of a Mathematics, Statistics or Data Science degree, Masters students in Statistics, Machine Learning and Management Science or Operations Research and of course first year PhD students. But most of all we hope it will be useful to those working as statisticians and data scientists in industry, commerce and government, and that it will provide sufficient motivation and interest to express some of their structural understanding of a problem within the simple but powerful framework of event trees and *Chain Event Graphs*.

The early development of these models was with Paul Anderson, Robert Cowell, Eva Riccomagno and Peter Thwaites. The latter two researchers have contributed and continue to contribute enormously to the development of the formal graphical and algebraic expression of Chain Event Graphs together with the development of causal inference techniques. Meanwhile, encouraged by detailed discussions with Henry Wynn and building on this early development, Christiane Görgen has discovered how to express and analyse the class of Chain Event Graph models formally as a family of statistical models. In the process of doing this, she embedded our methodology in the framework of graph theory, computer science, causal inference and what is known as Algebraic Statistics. This is outlined in Chapters 3, 4 and 8. Within the model selection technologies presented in Chapters 5 to 7, Guy Freeman first developed theory which was applied and enhanced by Lorna Barclay and Jane Hutton to develop new statistical methods whilst studying important Public Health applications of Chain Event Graphs. Simultaneously, Robert Cowell and Rodrigo Collazo developed and extended methods of model search to make the use of Chain Event Graph models feasible for the practitioner. Since that time a large team of other researchers has made serious theoretical advances with Manuele Leonelli, in dynamic versions of these technologies with Ann Nicholson and now a quite detailed causal theory, as well as recent new developments in decision theory and game theory. We do not have room to discuss all of these important developments properly but have cited them in the text. In particular, we hope to have given sufficient background to make the original articles accessible.

It is not possible to thank everyone we are indebted to here and who have helped us in this enterprise. However, throughout this development we have had detailed conversations with Martine Barons, James Cussens, Philip Dawid, Vanessa Didelez, Simon French, Paul Garthwaite, Hugo Maruri-Aguilar, Anjali Mazumder, Steffen Lauritzen, Catriona Queen, Nuala Sheehan, Milan Studený, Bernd Sturmfels and Piotr Zwiernik. All these researchers and many others have contributed directly and indirectly to this work. We wholeheartedly thank them all.

Rodrigo A. Collazo was supported by the Brazilian Navy and by CNPq-Brazil with grant number 229058/2013-2 for the period he co-wrote this book. Christiane Görgen was supported by the EPSRC grant EP/L505110/1 and was partly supported through the programme 'Oberwolfach Leibniz Fellows' by the Mathematisches Forschungsinstitut Oberwolfach in 2017. Jim Smith was partly supported by the EPSRC grant EP/K039628/1. Part of this work was supported by The Alan Turing Institute under the EPSRC grant EP/N510129/1.

List of Figures

List of Tables

Symbols and abbreviations

BN	Bayesian network
DAG	directed acyclic graph
f	density appearing in Bayes' Rule
α	hyperparameter of a Dirichlet distribution
$\overline{\alpha}$	equivalent sample size parameter
MAP	maximum a posteriori
BF, lpBF	(log-posterior) Bayes Factor
p, $p_{\boldsymbol{\theta}}$	(parametrised) probability distribution
P, $P_{\boldsymbol{\theta}}$	(parametrised) probability measure
Ω	finite and discrete set of atoms
n	number of atoms in a space
Δ_{n-1}	probability simplex of dimension $n-1$
CEG	Chain Event Graph
$\mathcal{T} = (V, E)$	graph of an event tree with vertex set V and edge set $E \subseteq V \times V$
v	vertex in a graph
u	stage in a graph
w	position in a graph
v_0, w_0	the root (position) in a graph
w_∞	the sink position in a graph
$(\mathcal{T}, \boldsymbol{\theta}_\mathcal{T})$	probability tree with graph $\mathcal{T}$ and a vector of edge labels $\boldsymbol{\theta}_\mathcal{T}$
$(\mathcal{C}(\mathcal{T}), \boldsymbol{\theta}_\mathcal{T})$	CEG $\mathcal{C}$ with underlying staged tree $(\mathcal{T}, \boldsymbol{\theta}_\mathcal{T})$
λ	a root-to-leaf/sink path in a graph
$\Lambda(\mathcal{T})$, $\Lambda(\mathcal{C})$	the set of all root-to-leaf/sink paths in $\mathcal{T}$ or $\mathcal{C}$
$\boldsymbol{M}_\mathcal{T}$, $\boldsymbol{M}_\mathcal{C}$	statistical model represented by $\mathcal{T}$ or $\mathcal{C}$
SCEG	stratified Chain Event Graph
$\boldsymbol{X}$	vector of random variables
$\boldsymbol{X}(I)$	vector of random variables following a particular ordering I
$\boldsymbol{\mathcal{X}}$	unordered set of random variables
m	number of random variables
$\boldsymbol{C}$	a class of CEG models
$\boldsymbol{C}(\Omega)$	the class of all CEG models on the same set Ω
$\boldsymbol{C}(\boldsymbol{X}(I))$	the class of all $\boldsymbol{X}(I)$-compatible CEGs
$\boldsymbol{C}(\boldsymbol{\mathcal{X}})$	the class of all $\boldsymbol{\mathcal{X}}$-compatible CEGs

$\boldsymbol{\mathcal{B}}$	block ordering of the set $\boldsymbol{\mathcal{X}}$
$q(\mathcal{C})$	prior probability of the model represented by $\mathcal{C}$
$Q(\mathcal{C})$	score of the model represented by $\mathcal{C}$
NLP, pmNLP	(pairwise-moment) non-local prior
AHC	agglomerative hierarchical clustering
DP	dynamic programming
CHDS	Christchurch Health and Development Study

1

Introduction

What is said is never quite what was thought, and what is heard is never quite what was said.

—Kevin Powers 'The Yellow Birds' (2012)

1.1 Some motivation

This book arose out of a long experience in the elicitation of expert judgements. A domain expert would begin by describing a series of events, often in natural language, and explain *what* might happen, *why* certain events might happen and if so then *how* these happen. A statistician would need faithfully to transform these verbal descriptions into a more formal probabilistic model. This model could be used to enable the domain expert to accurately express her understanding of a problem and to analyse its consequences. Furthermore, the probabilistic framework could be used to incorporate experimental evidence in a transparent, compelling and appropriate way. This would support the appropriate scoring of the efficacy of different options a decision maker has at hand using those expert judgements as her own.

Now anyone who has seriously engaged in the translation from a verbal explanation of a process into a probabilistic model knows that this is a highly sensitive task. This is especially the case when—as is often found—there are many different components to a problem that need to be accommodated into a probabilistic description. Thus there has been an increasing need for methodological frameworks which can help structure this translation. Perhaps the most important such structure is the *Bayesian network*—throughout for short abbreviated to the *BN* . We will introduce BN models formally in Chapter 2. Interestingly, some of the early motivation of the development of these networks was as an elegant and efficient way of expressing and processing information that would otherwise be expressible as a probability tree: see for instance [89].

BN technology has now been successfully applied to tens of thousands of applications—some extremely large. The BN modelling framework is an ideal one to use when:

1. The domain expert's description is concerned with the relationships that exist between a set of predefined measurement variables—variables that are then transformed into the random vectors defining a probability model.
2. Expert judgements are concerned with which of these variables might be useful to predict which other variables.

Such information can often be coded in a uniquely and formally defined acyclic directed graph (often a *DAG* for short) whose vertices represent the problem random variables and whose edges represent the existence of predictive relationships between these. This framework is sufficiently formal to be seen as expressing collections of conditional independence relationships in a probability model over the joint space of random variables. The graph can thus be used to code an expert's quantitative judgements about the extent of any relationships between the specified variables through a number of separate local conditional probability distributions. These then provide a well defined unique and coherent probability model for the whole system.

Such an incrementally elicited probability model will always be consistent with the global relational statements elicited from the expert's original verbal description of her problem: the most securely transferred knowledge from the expert. The less secure transfer of numerical evaluations of extent of relationships between variables is at least then mitigated through experts for various localised quantifications. Many pieces of software are available to support this process and the BN technology has thus made an important contribution to the process of probabilistic model building. We will devote much of the next chapter to describing in some detail how this process works and the useful methodologies it supports.

However, anyone experienced in using the BN framework will have also discovered that for many elicited problem descriptions this method is not necessarily an easy, natural or even feasible one to use.

The first issue concerns the first point made above. To use a BN it is first necessary to elicit a set of distinct measurement variables to capture a verbal explanation of dependence. Of course if the expert is simply describing how she thinks a set of given vectors of observations might exhibit a relationship or not then the set of measurement variables is obvious. But unfortunately the natural way of explaining a process may well not be expressed in such a way. Often instead the expert will choose to begin by describing various hypotheses about how various situations might unfold through the historical development of a unit in the process. For example, a doctor may begin by describing the possible developments a patient will experience after contracting an infection, the types of environmental exposures that might exacerbate the condition and the sequence of potential treatments that might be applied to address the illness and their potential side effects. An expert describing how a company launches a new product will often describe the various stages of the launch and the potential options it can take as the market receives it. An algae may undergo various stages of development that impinge on the ecology of a region and the different possible outcomes traced these organisms might develop over a season. A prosecuting barrister may describe the sequence of events he believed led to a suspect committing a crime. In all these diverse domains of explanations, the domain expert starts with a description of a *process*. Only later are measurements variables constructed around such explanations. Sometimes this is impossible to do in a credible or acceptable way. If the potential unfoldings of events are very different then the variables that explain developments over one branch may be quite different from those describing the development of another. The need to consider the values of a particular variable irrespective of the previous history of a unit in the population the model describes can even look absurd. For example in a surgical problem if an infection has already resulted in a patient's death how can we usefully talk about her temperature! Yet when building a system using a BN—or indeed any other of the standard graphical models—we would need to assign a probability to a number related to this pair in the conditional probability tables that would eventually populate such a model.

So the question is: within a discrete system can we represent these possible developments that a unit might experience directly in some sort of coherent mathematical framework supported by a graph? The answer is of course that we can. For centuries the event tree —a graph formally defined in Section 3.1— has done exactly this. In fact, trees are widely used in many disciplines, especially in computer science, to depict the structure of a process. They are also heavily used in statistics as a descriptor for classification, in decision theory to represent decision rules and they have increasingly been formally studied by probabilists as objects of profound formal interest in terms of stochastic event spaces. Hence there is at least one structural alternative to standard classes of graphical DAG models when modelling the evolution of discrete problems.

Just as with a BN we can first use a directed event tree graph to describe a problem. Only afterwards do we assign probabilities to all its atoms—depicted as root-to-leaf paths—and thus fully populate this elicited event tree into a probability tree. This can be done coherently simply by adding edge probabilities to the event tree noting the probabilities on its atoms are simply the product of these delivered (conditional) probabilities.

This book and the technological developments it reports were inspired by the early content of a book by Shafer [90] and subsequent conversations with its author. Shafer argued that event trees are not only a ubiquitous practical tool but that they also have a profound representational power. In particular, he argued that these representations are at the root of the best expressions of causal hypotheses. Families of probability trees thus provided a framework for moving from an explanation of a process to a probability model which is philosophically sound. Such trees could therefore be used as a very flexible and expressive inferential framework to express complex and multifaceted problems of inference to a great level of generality—far greater than that provided by other existing tools like BNs. So certainly the probability tree should never have been relegated to be simply a vehicle to express some simple concepts in high school probability classes as had been the case until about a decade ago.

One reason the event tree has not been used as widely as it might is that the development supporting architecture for elicitation, interpretation, inference, model selection and causal embeddings had only been patchy. Over the last ten years teams of researchers, many based at the University of Warwick, have developed methods that systematically and consistently have addressed these issues and proposed solutions to these. By adding colours to the edges and vertices of an event tree and transforming its graph we find that we can obtain a new graph that can be used just like a BN. We call this new graph a *Chain Event Graph* or for short a *CEG*: see Section 3.1 . CEGs can be used to make early inferences about a system without first committing to numerical values—but through the colourings we use—and by building on statements about when the unfoldings of various situations are driven by the same probabilistic process and when they might differ. We shall see that such statements are actually disguised (context-specific) conditional independence statements . The methods developed so far have been mainly Bayesian but we view this as a stylistic choice. All the methods we describe below for providing a comprehensive methodological environment for Chain Event Graphs have their non-Bayesian analogues. We hope to see many such developments in the future.

This book therefore aims to bring to a wide audience an encompassing picture of the promise of this new class of CEG models which can be used for analysing many common classes of problems in a coherent, comprehensive and transparent manner. We demonstrate how we elicit event trees by using descriptions about how things might unfold in time or explanations about what might happen under different scenarios. Using the tree and asking about various symmetries in its associated probabilities then enables us systematically to recast these stories into families of statistical models. Thanks to this development, standard methods of analysis can then be used directly to make inferences about the described hypothesised underlying processes: see Chapters 4 to 6. In this way the tree becomes not only an instrument of elicitation but inference, decision and post-analysis explanation that justifies—directly in terms of the elicited framework—why certain inferences can be made and why certain decisions can be taken. Specifically, collections of statistical hypotheses associated with competing unfolding stories can then be formally examined for their evidential support and the method be used to help guide future action.

1.2 Why event trees?

Event trees have a long and profound history dating back at least to the philosopher David Hume [90]. With the formal development of discrete probability theory the term 'event' was defined to mean simply any subset of atoms that was part of a sigma field. However, in common language and its earlier meaning an event was a much richer construction. It typically signified something that happened or might happen in the development of a scenario, person or object of study. In fact, this usage is still preserved in some areas of statistics—for example within the term 'Event History Analysis'. However, through developments made by the Russian school of statistics, probability theory began to be treated as an elegant axiomatic system. For the finite discrete problems we consider here the term 'event' was then used simply to define a set of atoms in a space. The concept of an event as a *component in a description* of how circumstances unfolded and as part of a description which is dynamic was lost. First probability courses are still taught in this way and the static object of an event as defined by the Russian school still appears in most elementary probability texts.

With Shafer [90] we would like to contend that it is often helpful to think of an event as part of a narrative about how the past might have evolved or the future will evolve. The event tree is the natural vehicle to express this more flexible interpretation of the term. Event trees and probability trees will be familiar constructions to many readers. Before we formally pin down precise definitions in Chapter 3, we can first introduce some of the modelling ideas less formally, through two illustrations which will hopefully help to motivate the later development in this book.

A first example from forensic science

We start with a very simple example drawn from recent problems concerning the deductions and explanations that a forensic scientist might need to make and communicate in a court of law. One author was emersed in this type of application when commissioned to design probabilistic decision support systems for forensic scientists [81]. The following example is a simplification of possible outworkings of a rape case presented to a jury. This type of reconstruction is what is routinely delivered to a court for many serious crimes. Prosecution barristers deliver as compellingly as they can the case against a suspect and the defence barristers for the suspect in the dock. These arguments are often related as a story or collection of possible stories about how things might have happened. Jurors then need to assess whether the suspect is guilty of the alleged crime beyond all reasonable doubt or not.

The role of forensic scientists who need to be called to help jurors to properly assess the strength of the two cases in the light of any evidence is an increasingly delicate one. They need to carefully and professionally process the data from any evidence they have available and present their analysis as truthfully and as unambiguously as they can when called as an expert witness to a case. In particular, the forensic scientist needs to communicate evidence in such a way that jurors can appreciate the forensic evidence within the context of the stories told by the different barristers. With the advent of new technology there is increasing pressure for forensic scientists to need to analyse and sometimes appropriately combine a variety of different sources of evidence and make valid and defensible deductions about the strength of evidence behind these sources.

The Bayesian network is increasingly being used as an aid to help forensic scientists come to a proper assessment of strength of evidence of this kind [103]. However, when the arguments of the defence and prosecution barristers tell very different stories about what had happened and various pieces of evidence only apply to a subset of these stories then event trees can often be a much more useful tool. These trees are often linked to 'activity level evidence' where the variables needed to be considered for the prosecution case are often different for those needed for the defence case. Pooling these two cases—as necessary when using a BN—is cumbersome and potentially confusing to all concerned. Even in this very simple criminal case we can illustrate this point.

Assume that all parties agree the following aspects of the unfolding events in a criminal assault are true. A woman wearing a recently washed dressing gown was attacked by an assailant at her home at around 11pm at night, assaulted and raped. She remembers nothing that can describe the assailant other than that he was male and had a full head of hair. However, a forensic analysis of the hair on the dressing gown the woman was wearing during the assault discovered exactly one hair on it which was not hers. Because of other aspects of the case, all agree that this hair must have been donated during the assault. The DNA signature of this hair was found to match that of suspect **S** who was subsequently arrested on this basis. All agree that the woman and **S** were strangers and there was no reason for them to have met previously or for **S** to be at the victim's house legitimately. So the woman could not have donated a hair from **S** herself in a more innocent circumstance. The prosecution contend that **S** assaulted the woman. The defence, on the other hand, claim that **S** was not the assailant and was not in the vicinity of the victim's house at the time of the assault. Hence the hair must have been donated by some other unknown person **U**.

In normal circumstances in coming to their decision the jurors are encouraged to consider that one of the possible unfoldings of events as described by the barristers is the true one. From a statistical point of view this means that they should assign zero probability to anything other than the stories presented by these barristers. Here the prosecution has a simple story to describe her version of events. The suspect was in the vicinity of the victim's house, broke in and assaulted and raped her. In the struggle he donated a hair on

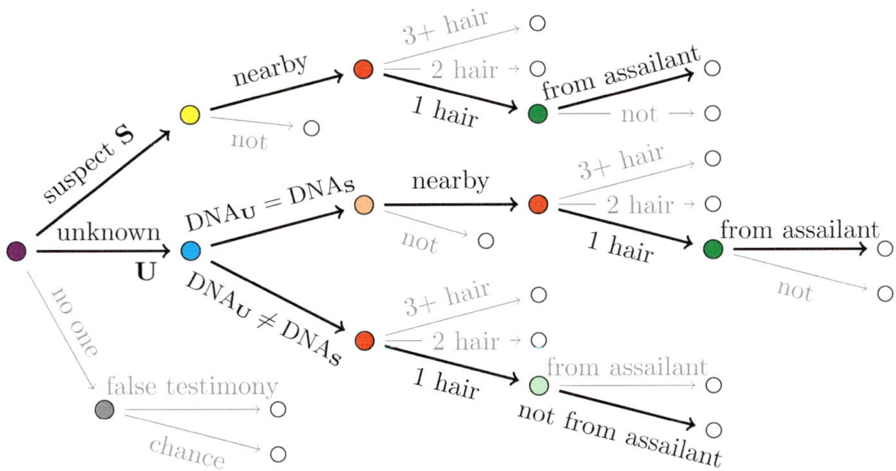

FIGURE 1.1
An event tree depicting possible unfoldings from the legal case example. Black depicted paths correspond to the defence and the prosecution case. Colours identify vertices with equal transition probabilities.

to her dressing gown. This hair—which was subsequently retrieved by those examining the case—came from the suspect.

On the other hand the defence said that someone else made the assault and therefore one of two other possible things must have happened. The first possibility was that the suspect was not in the vicinity of the house but an unknown other person **U** was. This person shared the same DNA profile of **S**. In this case this would explain the rest of the acknowledged events: the whole story would be the same except with **U** replacing **S**. This would explain that **S** was innocent. Alternatively, another possibility would be that **U** did not share the DNA profile of **S**. However after raping the victim he donated a hair from **S** (or someone else who shared **S**'s DNA profile). The defence barrister plans to contend in court that this is indeed a possibility because any individual can carry hairs on their clothes from other people as illustrated by the victim herself. It must have been this hair from **S** that **U** had previously picked up and donated to the dressing gown and not his own hair. That **U** carried a hair from **S** is all the more probable because **S** often engaged in fights with other men such that **S** could easily donate some of his hair on to their clothes. This provided a second explanation of **S**'s innocence in the light of the evidence.

All the relevant elements of this story can be represented using the event tree in Figure 1.1. Each path of this tree from its root to its leaves then provides a graphical description of the different stories in this legal case.

The single piece of forensic evidence is the single recovered hair that matches the DNA of the suspect. When we condition on this event we are left with only three root-to-leaf paths that can be assigned anything other than a zero probability. Graphically, this conditioning can be achieved by setting all probabilities associated to impossible unfoldings to zero and greying out all these edges and vertices.

From the usual rules of conditioning, the probability of lying on one of the paths with non-zero probability is simply proportional to the product of the conditional probabilities along the relevant edges. The proportionality constant is simply the reciprocal of the sum of these remaining atomic probabilities. The tree can thus be used to calculate conditional probabilities on the atoms of this event space resetting their atomic probabilities to zero and renormalising all the remaining atomic non-probabilities so that they sum to one.

We will now perform these operations explicitly on the case above. In this example the events that are relevant to the prosecution and defence cases are as follows:

- the event S that the suspect **S** committed the crime and the event U that some unknown man **U** committed the crime,
- the events $N_{\mathbf{S}}$ that **S** and $N_{\mathbf{U}}$ that **U** was nearby when the crime took place,
- the event H that exactly one hair from the assailant was retrieved from the victim,
- the event A that the hair retrieved from the dressing gown belonged to the assailant and not to anyone else, and
- the event D that the DNA of **S** and **U** match each other.

These events label the edges of the tree as shown. To specify a full joint probability model we now need to embellish this event tree with edge labels that are probabilities, thus turning it into a probability tree. These will be the relevant conditional probabilities a juror might entertain *before* the evidence is presented by the forensic scientist in the light of any background evidence. For a coherent probabilistic judgement to be made these conditional probabilities on edges are probabilities conditional on all the events preceding them in the narrative behind their respective root-to-leaf path in the tree.

In this example these edge probabilities are as follows:

- $P(S) = \sigma$, the probability that **S**—rather than **U**—committed the crime,
- $P(N_{\mathbf{S}}|S) = \nu_S$ the probability that **S** was in the vicinity at the time of the crime,
- $P(N_{\mathbf{U}}|U, D) = \nu_U$ the probability that someone else **U** who committed the crime and shared **S**'s DNA was in the vicinity at the time of the crime,
- $P(D|U) = \delta$ the probability that another unknown man **U** shared the suspect's DNA,
- $P(H) = \theta$ the probability any assailant would leave exactly one hair on the dressing gown of the victim, and

- $P(A) = \alpha$ the probability that a hair donated to someone in an assault is their own.

If **S** had a brother with unknown DNA signature who could be another suspect then there would be a significant probability δ that this brother and **S** shared the same DNA. Similarly, a juror might be content to assume that the probability θ is likely to be the same whether the assailant was **S** or **U**, unless the suspect was unusually hairy or bald; we have assumed this in the original description of the case above. Again with this caveat we might assume α to be the same whoever made the assault. We embed these assumptions by colouring those vertices in Figure 1.1 which have the same attached probabilities using the same respective colour. In Chapter 4, we will then provide a more detailed interpretation of this colouring: see Example 4.10 and in particular Figure 4.4.

The prosecution has simply one case so the probability of the suspect being guilty is given by the product $\sigma\nu_S\theta\alpha$ of edge labels along the root-to-leaf path in the tree corresponding to that story.

The defence case is given by two stories, represented by separate root-to-leaf paths. The probability that the suspect is innocent is therefore proportional to the the sum of the two products $(1-\sigma)(\delta\nu_U\theta\alpha + (1-\delta)\theta(1-\alpha))$.

Of course the key issue is how much more probable guilt is than innocence. This is equal to the ratio of the two probabilities calculated above:

$$\frac{P(\text{suspect guilty})}{P(\text{suspect innocent})} = \frac{\sigma\nu_S\theta\alpha}{(1-\sigma)(\delta\nu_U\theta\alpha + (1-\delta)\theta(1-\alpha)))}. \tag{1.1}$$

There are various points to make here.

First this critical ratio of probabilities is expressed simply in terms of a sum of products of conditional probabilities associated with the edges along one of the three extant root-to-leaf paths of the tree. These formulae not only respect the arguments of the barristers but also do this in a transparent whilst completely formal way. This rather hides the subtle features where we must get the conditioning events in the right order before we can assign the appropriate probabilities. Horrendous error can be made if we encourage the jurors to substitute the wrong probabilities in the wrong conditioning order—this is known as the *prosecutor's fallacy*, for instance in [97]. By closely following the ordering of events in the narrative and by representing this narrative using the tree we automatically guard against such errors.

Second both the prosecution and the defence edge probabilities neatly partition into two sets: the set of probabilities that must be the sole responsibility of the jurors and that must not be influenced by the scientist, and those where the assessments are scientific in nature and can properly be informed by forensic science. In particular, the probabilities in the first three items listed above must be in the domain only of the jurors. The last three items can properly be informed by evidence from a forensic scientist. For example, they can properly share with the court how common it would be for someone else to share **S**'s DNA within the circumstances of this crime. This would inform δ. Systematic well designed surveys have been conducted to sample the probability distribution of the transfer and subsequent retrieval of hair given activities like those reported in the crime and the forensic analyses of these experiments also legitimately inform both θ and α and can be shared with jurors.

This leads us to a third point. Calculations like those in (1.1) are functions of only two arguments, the tree graph and its possibly identified edge probabilities. The point is here that it is plausible to suppose that jurors will be happy with both the tree and its colourings as a faithful representation of the inferences they need to make. Our representation then has a generic integrity that allows us to make deductions about the relevance of various features of the problem and to hand jurors the *formulae* they need to assess the guilt of the suspect. Hereby the probabilities that jurors might assign to the final ratio (1.1) could legitimately diverge radically—each taking their own view on the plausibility of different elements of the case and even the credibility of the forensic scientists themselves. So this coloured graph is all that can be assumed to be shared amongst those making inferences of a generic kind.

Can thus a coloured tree only be used as a basis for depicting the shared structural information—that lead us to formulae like (1.1)—or perhaps also to preprocess this information so that life can be made as simple as possible for the jurors? The answer is that it can! In particular, we can work out from the coloured event tree and another graph constructed from it—the CEG—what might be and is not relevant to the case *before* it is brought to the jurors.

The graphical properties of BNs are consistently used for the task of determining which aspects of the problem need to be drawn into the argument for the inferences that need to be made. This can be done before any numerical calculations have been made. Completely analogously, the CEG can be used to make these sorts of inferences by for example considering its 'cuts': see Section 4.1. A cut is a set of vertices through which all root-to-leaf paths must pass exactly once. We notice in the event tree in Figure 1.1 that all black paths pass through situations coloured red—these are associated to the event H of finding precisely one hair. Thus these red vertices form a cut. We can now read off the independence statement 'innocence or guilt of our suspect does not depend on θ'. In Example 4.10, we will both give a CEG representation for the event tree discussed here and outline the rationale behind this conclusion. Since in (1.1) the probability θ cancels out this is indeed true! In this sense the fact that exactly one hair was found not belonging to the victim should be quite irrelevant to any juror's deliberation: so the introduction of the probability in the presentation by an expert witness could only introduce confusion. In fact, the forensic scientist would know that in any similar case this feature of the evidence would not be relevant and so include this into any protocol used for such cases. Much less transparent deductions can be made in moderately sized rather than small problems like the one above. This can dramatically simplify an analysis or explain the generic dependence structure of the problem.

Fourth we observe that in our example various elements associated with the defence case—for example the possibility that someone else shared the suspect's DNA—do not come into the prosecution case. Usually in court cases this sort of asymmetry is very common. Because the usual formulation of these issues through a BN requires that we need to introduce the same collection of measurements to describe both sides of the case, the methodology surrounding DAG models is not entirely suited to use in the sorts of settings where 'activity level' evidence is being considered. But the coloured tree and Chain Event Graph are!

Fifth note that if subsequent to the initial deliberations the court is asked to contemplate new possible explanations of what might have happened then these can simply be added to the tree. The new coloured event tree can then be adapted while holding much of its structure the same to investigate how the evidence might inform what might have happened. For example, there might be new information that the incident did not happen at all, that the suspect was actually in custody at the time of the assault or the victim had worn the dressing gown recently whilst staying with a friend after a party held there. Each of these new scenarios could be introduced by embellishing the event tree and adapting its colouring. The reader will be asked to do this in Exercise 1.2 below. This process would be much more difficult when using a BN!

Sixth, this whole analysis has been conducted purely using the structure of the event tree. The big advantage of this is that each barrister can read off a narrative which explains why their case should be supported for the figure given.

Finally, once the analysis has taken place, we can then also formally build into this analysis any uncertainty a juror might have about the probabilities delivered by the forensic scientists. Thus these probabilities can be treated in a fully Bayesian way: being considered themselves as random variables. The way this is done for individual probabilities is described in the next chapter and more specifically for edge probabilities in a tree in Chapter 5. These assessments can then be pasted together to allow formal account to be taken of the robustness of any inferences about a suspect's guilt or innocence within the delivered forensic evidence.

Typically of course in such cases the possibilities as part of the prosecution and defence case can be much more numerous and the inferences made from it much more subtle. But then the technologies we describe in this book become more and more powerful and less and less routine.

Of course all the techniques illustrated above can be used outside the court, from personalised medical diagnoses, to historical reconstructions, to future predictions of market perturbations and scenario analyses. Any problem that can be described in terms of a number of different possible trains of events can be used in the way we have illustrated.

1.3 Using event trees to describe populations

In the previous section we have only concerned ourselves with inferences about a single individual or a single outworked scenario. However graphical methods are also widely used to study the behaviour of populations each of whose members is assumed to respect the probability description expressed by a shared graph. Coloured trees as in Figure 1.1 can be used for this purpose as well and have indeed now been successfully employed in a number of applications that we present throughout this book.

One common use of standard graphical models is to perform an exploratory data analysis. Here the point is to summarise and structure the data available to a client in a way that might make sense to her and help her to discover new hypotheses to investigate. For this purpose models are selected from a class which appears at first sight to best represent the data seen in an experiment survey, observational study or a combination of these forms of empirical information. In Chapter 5 of this book we show how such a model search can be performed over classes of CEGs to find the CEG which seems to best describe the data generating process underlying empirical information. Within the Bayesian paradigm perhaps the most popular search is Bayes Factor MAP search which selects the model that roughly speaking has the highest probability among those searched: see Section 2.1. This and other standard BN model selection techniques can be adapted so that they apply equally well to different classes of CEGs which can then be estimated and scored in closed form. One of the key advantages of our new models is that they can select explanations that support a narrative which may or may not be consistent with some BN. Because the CEG is a much richer class of models than the BN which it contains as a simple special case, we often find that for any given dataset these explanatory CEG models explain the data *much* better than any competing DAG model: see our in-depth analysis of another more substantial real-world dataset in Chapter 7.

An example from health science

We briefly illustrate how the methodology we developed for coloured event trees and CEGs has been applied to better understand and predict life expectancy of sufferers of cerebral palsy. A detailed discussion of the full study from which this example is drawn is given in [6]. The problem is here largely simplified by using only those patients with full records and considering only three explanatory variables. The first, ‘birth weight’, can take three levels

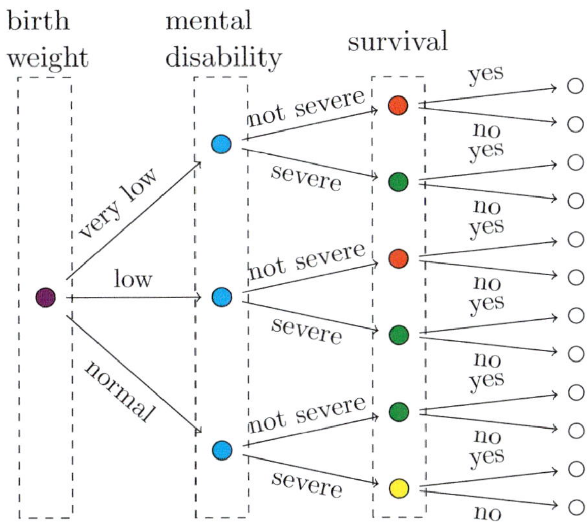

FIGURE 1.2
The highest scoring event tree inferred from a cerebral palsy dataset analysed in [6].

whilst the other two variables, 'mental disability' and 'survival', are indicator variables and so can take one of two levels. An event tree for representing these variables can take the form given in Figure 1.2.

We investigate the following issue: given the evidence of the data available, is it possible to plausibly equate some of the edge probabilities with each other and thus simplify the tree? The only graphs that might make sense in this context are the ones that allow us to assert whether on the basis of the data the vertices associated with mental disability vertices can be identified with each other, and which survival vertices can be identified in the same way. As reported in the publication above, we found that the best scoring model equated all the pairs of mental disability edge probabilities, and partitioned the six survival vertices into three sets: each containing equal pairs of edge probabilities.

The highest scoring coloured tree was therefore the one given in Figure 1.2 above. Its corresponding CEG will be given in Chapter 4 where we present a more detailed discussion of this problem: see Example 4.11 and in particular Figure 4.5.

There are again various points to notice here.

The first point is that the tree we found above does not correspond to a BN in the sense that it could not be represented faithfully by a single DAG. This supports our choice of making this more elaborate model search across the data! In fact the best fitting model as represented by the coloured tree is actually a context-specific BN: see [62] and Chapter 3.

The second point to notice is that the discovered graph can be directly read back to the client. In particular, reading from the top of Figure 1.2 we can conveniently partition surviving patients into three risk classes. Using the estimated survival edge probabilities given in the paper from which this example is drawn, the highest rate of survival was associated with patients who had low or very low birth weights and no severe mental disability. The middle survival group consisted of those with low or very low birth weights and severe mental impairment, or normal birth weight and no severe disability. The patients with the worst survival rate and at most risk were those with a normal birth weight but severe mental disability. The other point we can draw from the remnant colouring in the tree is that in the study population there was no strong evidence that birth weight and mental disability were related to each other.

Third, because the search space of coloured event trees and CEGs is huge compared with that of the BN for even moderately sized problems, there are significant technical challenges in searching the whole space. However, just like above there is often simple domain information which can help to vastly reduce the number of explanatory models. Furthermore formal search methods can be devised which force explanatory models to be as simple as possible. An extensive discussion of some of these issues is given in Chapter 6.

Finally, the full dataset for the example above contained many patients whose records were not complete. For most graphical models this significantly complicates any search because this missingness draws us out of the model class—even more so when the missingness is informative. This need not be the case for CEGs. For example the setting above which also contains all patients with some of their records missing can be searched in exactly the same way as the complete data: see Section 5.1.

1.4 How we have arranged the material in this book

In Chapter 2 of this book we will introduce the background material on which this work is based. This will concern various ideas about discrete statistical modelling with a focus on Bayesian inference. In particular, we will outline central results from the theory of BNs, repeating very briefly how this popular class of models has been used in the past. By studying BNs we implicitly present properties of a very well studied subclass of CEG models. Of course this cannot do justice to the vast field of BN modelling. However, we refer to more detailed developments of this area throughout. In later chapters we will further use these results to design many analogous methods for the statistical analysis of the larger class of CEGs.

These new graphical models are first introduced in Chapter 3. We will start our development from event trees, equip these with edge labels and a probability distribution and then define parametric statistical models represented by this type of graph. In this powerful framework, properties of the underlying probability distributions can be coded using a colouring of the graph. This then enables us to build the CEG as an equivalent and advantageous representation of coloured tree models. Various examples will explain the semantics of these graphs and their links to decision trees and other tree-based methods used in statistical inference and decision analysis.

Chapter 4 opens the door for the use of CEGs in probabilistic inference. Here, we explain how to embed a given problem description into a Chain Event Graph and how to, vice versa, read modelling assumptions directly from a CEG. Conditioning operations, graph-defined random variables and separation theorems are briefly presented and widely illustrated. We then move on to exploring the properties of discrete and parametric statistical models represented by a CEG. These models can be depicted as hypersurfaces, curves or other subsets in high-dimensional probability simplicies. In particular, we can specify them as the solution sets of collections of odds-ratio equations. We then analyse when two different CEGs represent the same model—a property called statistical equivalence. For this purpose we define two local graphical operators, called a swap and a resize, which are easily handled and enable the modeller to elegantly traverse such an equivalence class. These operators are then employed in the analysis of a simplification of a real system.

In Chapter 5 we present how to use CEGs in statistical inference. In particular, it is well known that BNs support a conjugate analysis using prior distributions which are product Dirichlet distributions. These enable the modeller to seamlessly move from probabilities expressed through subjective expert judgements to situations where data informs these judgements and can easily be accommodated into the same model. We show how this technology transfers more generally to allow conjugate inference over the parameters of a CEG. With these results in place, we then show how a Chain Event Graph can guide the quick calculation of conditional probabilities of interest given new information even when the underlying model is large. The propagation algorithm we present here for calculating these probabilities is both more flexible and often faster than its BN analogue.

To set up a CEG in practice it is necessary to elicit probability distributions over its parameters. In Chapter 6 we show how to do this. In the common situation that there are many different candidate explanations—each expressed as a CEG—we show how to find the one which might best explain a given scenario. In particular, we present two popular methods of model selection: using maximum a posteriori selection methods and non-local priors.

Then in Chapter 7 all threads of our methodological development are drawn together when we apply the various results of the previous chapters in the thorough analysis of a real-world dataset. This analysis is based on a thirty-year longitudinal study following a cohort of over one thousand children in New Zealand, analysing factors that drive child illness. We bring our technical results to life in presenting eight competing CEG models describing this dataset and comparing them—in score and in terms of the setting—to results from previous studies and to three competing BN models.

Chapter 8 is the most self-contained chapter of this book. It rounds off our analysis by giving a sneak preview into ongoing work in using CEGs as viable tools in causal inference. We illustrate here how interventions in a system can be translated into operations on a given graph and outline the advantages of this depiction over standard BN methodology. The notion of a 'causal' CEG is defined and analysed in a real setting drawn from the preceding chapter. Some simple ways of demonstrating how CEGs can be used for causal discovery are presented.

1.5 Exercises

Exercise 1.1. *Consider your work schedule tomorrow. Write down the possible outworkings of your day as a tree. Identify the edges in that tree whose probabilities you are happy to assert are the same.*

Exercise 1.2. *In the legal case above depict the three different event tree descriptions that you would obtain if*

1. *the defence barrister wanted to assert that the event described by the victim never happened, or*
2. *that the suspect was known to be in police custody at the time of the assault, or*
3. *that new evidence said that the dressing gown had been recently worn elsewhere at home where a party had been held.*

How might these change the probability of events and which probabilities can still be identified as equal to each other?

Exercise 1.3. *From the tree in Figure 1.1 construct an argument in English to present to the court first for the prosecution and then for the defence.*

Exercise 1.4. *In the cerebral palsy example carefully decompose the meaning of the coloured tree given in Figure 1.2.*

2

Bayesian inference using graphs

This chapter will give a very abbreviated overview of the vast field of statistical inference. We focus in particular on modelling from a Bayesian perspective and the use of graphical models. The methods we present here can of course, in the shortness of space, only be illustrated in very small scale examples and cannot do justice to the very detailed discussions provided by books which focus on these topics exclusively [22, 63, 77, 97]. However, we will present below some of the key ideas and results in this field: we start from Bayesian inference in any discrete model and then move on to discuss conditional independence models represented by directed acyclic graphs (DAGs). Throughout, these models will be called Bayesian networks (BNs).

We will then very briefly discuss inference in BNs, graphical results like the d-separation theorem, model selection and we will introduce propagation algorithms. This is so that over the course of this book we will be able to present a large amount of ideas which have been developed as extensions or parallel developments to these concepts for CEG models. The notation we set up over the course of this chapter will then be used as a reference.

2.1 Inference on discrete statistical models

When setting about the task of building a probability model we are faced with a number of questions. The first concerns exactly what the probability model is about and requires us to choose an appropriate event space. In this book we will discuss only finite discrete probability models since the Chain Event Graph is always one of these. So in this setting our question can then be rephrased as: *What is the set* $\Omega = \{\omega_1, \omega_2, \ldots \omega_n\}$ *of mutually exclusive outcomes for each unit in the modelled population and which is sufficiently refined to enable us to answer all the questions we want to ask?*

Each of the elements $\omega \in \Omega$ of that space is also called an *atom*. Throughout this book we will assume that Ω is known and that each of its atoms has a distinct meaning. For instance, if $\Omega = \{\omega_0, \omega_1\}$ then ω_1 might denote that a particular unit was infected with a virus and ω_0 that she was not. We henceforth assume that we all agree what this statement 'infected by that virus' exactly means. We often have to be very careful in practice to ensure that this is so! We also of course need to know that ω_1 is the atomic event that the unit has the infection and ω_0 the event she does not—and not the other way around. Thus Ω is determined by the model setting. In particular, it is usually wise to choose $n \in \mathbb{N}$ such that Ω is the smallest set of outcomes that have a relevance—directly or indirectly—to the questions being asked of the domain. The modelled population might hereby be very large or simply—as for example in the legal example of the previous chapter—concern only one person.

Once we have determined the appropriate set of atomic events, the second question that needs to be asked is: *How do we assign probabilities* $p(\omega)$ *to each of the atoms* $\omega \in \Omega$*?* Equivalently, what we need to specify is a probability mass function $p(\omega_i)$ for all $i = 1, 2, \ldots, n$ where of course—by the usual rules of probability—all values are positive $p(\omega_i) \geq 0$, $i = 1, 2, \ldots, n$, and sum to one, $\sum_{i=1}^{n} p(\omega_i) = 1$.

There are two direct ways to obtain such a function. The first is to use elicitation techniques where a group of domain experts is charged with the task of specifying these numbers. This can be done for instance using scoring rules or gambling preferences [74]. Sometimes we have no alternative but to specify probabilities in this way. However, this method is fragile and susceptible to many biases in the elicitation process. Furthermore, when there are more than a few atoms in the space the elicitation process can be extremely time consuming and the inevitable number of small probabilities are then only accessible with considerable elicitation error. A careful discussion of these issues can be found in [35, 42, 97].

Alternatively, when we have a random sample of size $N \in \mathbb{N}$ of the population we might instead use the corresponding sample proportions $\hat{p}(\omega)$—the *maximum likelihood estimate*—we observe within that sample at each of the atoms $\omega \in \Omega$ in order to estimate each probability $p(\omega)$. This can work very nicely if N is massive compared with the number of atoms n. Sadly this is very often not the case. Instead the number of possibilities n we want to distinguish is usually very large and typically random sampling is expensive so that N is often only moderate in size. Then this method does not work well. To illustrate this consider the example discussed above where we randomly sample two people from a population of individuals susceptible to a viral infection and discover that neither have the infection. Using sample proportions we would assign probability 1 to no-one in the population having the infection. This surely is not sensible! In general, this method will assign zero probability to anything not seen in the sample. This of course is very myopic unless N is so large compared with n that all estimates are non-zero, so $\hat{p}(\omega) \neq 0$ for all atoms $\omega \in \Omega$.

Even when we have 'big data' so that N is extremely large we will also have the difficulty that we have not sampled randomly so that we cannot automatically assume that the mass function of a unit from the sampled population is the same as the probability mass function associated with the whole population. For example, consider a crowd sourced sample where people are asked to report whether they have seen a particular insect on a given day in a given region of the country. These reported sample proportions obviously cannot be directly used to estimate the proportions of these insects. In particular, people will report these events only if the weather conditions at that place are conducive to both humans and the insects being out. We also know that rarer insects are more likely to be reported because seeing them is more of a prize for the observer. These issues can introduce very strong biases that make naïve methods of estimation like sampling proportions inappropriate. To use such data we need to process the information using some kind of statistical model *before* we can make use of the data we observe.

To overcome these difficulties statisticians have devised ways of processing information and combining these with expert judgements to come to better ways of estimating probabilities. From a Bayesian perspective, this is done by first building a joint probability model of what we might observe $\boldsymbol{Y} = \boldsymbol{y}$ within a sample and the unknown vector of probabilities $\boldsymbol{p} = (p(\omega_1), p(\omega_2), \ldots, p(\omega_n))$ whose entries are called *atomic probabilities.* In this context, $\boldsymbol{p}$ can also be thought of as the vector of values a random vector might take. The joint probability model can then be constructed in two steps:

1. We specify, for each $\boldsymbol{p}$, the probability mass function $f(\boldsymbol{y}|\boldsymbol{p})$ of the events $\boldsymbol{Y} = \boldsymbol{y}$ we might observe *conditional* on each of the possible different values of $\boldsymbol{p}$. This is particularly simple if what we observe is a random sample of the whole population: see below. Sometimes, $f(\boldsymbol{y}|\boldsymbol{p})$ is called the *observed likelihood* of the parameter $\boldsymbol{p}$ given the data $\boldsymbol{y}$.
2. We then specify the marginal probability density $f(\boldsymbol{p})$ of the probabilities $\boldsymbol{p} = (p(\omega_1), p(\omega_2), \ldots, p(\omega_n))$. This joint density—called our *prior density on* $\boldsymbol{p}$—will reflect any expert judgements we might be able to bring to the inference.

Then by using Bayes' Rule

$$f(\boldsymbol{y}, \boldsymbol{p}) = f(\boldsymbol{y}|\boldsymbol{p})f(\boldsymbol{p}) = f(\boldsymbol{p}|\boldsymbol{y})f(\boldsymbol{y}) \tag{2.1}$$

we can in particular calculate our *posterior density* $f(\boldsymbol{p}|\boldsymbol{y})$ of our vector of atomic probabilities $\boldsymbol{p}$ given what we have seen in our sample, $\boldsymbol{Y} = \boldsymbol{y}$. Because this is a density and not just a single marker like a sample proportion, it gives us not only a set of good estimates of the actual values the vector of probabilities might take but also measures of how certain we are about that estimate, via for example the variances of the different components of this vector. In this context $f(\boldsymbol{y})$—called the *marginal likelihood* of $\boldsymbol{y}$—is the probability based on our expert judgements we would a priori have assigned to the data we actually observed. Thus this term is a natural one on which to base a score of the success of our model in explaining the data we subsequently saw. We refer the reader to [95] for an introduction to Bayesian inference.

2.1.1 Two common sampling mass functions

One of the simplest but surprisingly common sampling models is the binomial sampling model. Here, as in the viral example above, the sample space $\Omega = \{\omega_0, \omega_1\}$ has just two states corresponding to the absence or presence of a feature of interest—there the virus—within the population. Suppose we simply sample N of the units in that population and count the number y of those units with the feature, so that $N - y$ do not have the feature. Then standard probability theory tells us that for $y = 0, 1, 2, \ldots, N$, we obtain a probability density

$$f(y|\boldsymbol{p}) = \binom{N}{y} p(\omega_0)^{N-y} p(\omega_1)^y \tag{2.2}$$

where $\boldsymbol{p} = (p(\omega_0), p(\omega_1))$ denotes the probability vector with components $p(\omega_0), p(\omega_1) \geq 0$ which are positive and sum to one, $p(\omega_0) + p(\omega_1) = 1$. Hence Y has a *binomial distribution* $\text{Bin}(N, p(\omega_1))$ with parameters N and $p(\omega_1)$.

The more general case where the sample space $\Omega = \{\omega_1, \omega_2, \ldots, \omega_n\}$ has $n > 2$ atoms is hardly more difficult. Thus suppose each unit in a randomly sampled population of N could take one of n features in the sample and that y_i units in the sample are assigned to that feature ω_i. Then the sampled proportions on these atoms are simply given by the fraction $\hat{p}(\omega_i) = y_i/n$, $i = 1, \ldots, n$, just like in our discussion of maximum likelihood estimation in the previous section.

We henceforth write $\boldsymbol{y} = (y_1, y_2, \ldots, y_n)$ for the vector of counts, $\sum_{i=1}^{n} y_i = N$. Then the random variable $\boldsymbol{Y}$ must have a *multinomial* $\text{Multi}(N, \boldsymbol{p})$ distribution conditional on the values of the vector $\boldsymbol{p}$ of atomic probabilities. Here the probability density is given by

$$f(\boldsymbol{y}|\boldsymbol{p}) = \frac{N!}{y_1!y_2!\ldots y_n!} p(\omega_1)^{y_1} p(\omega_2)^{y_2} \cdots p(\omega_n)^{y_n} \tag{2.3}$$

where again, being a vector of probabilities, $\boldsymbol{p} = (p(\omega_0), \ldots, p(\omega_n))$ is such that $p(\omega_1), p(\omega_2), \ldots, p(\omega_n) \geq 0$ and $\sum_{i=1}^{n} p(\omega_i) = 1$.

Remarkably, we find that for all the issues we need to address in this book, all our problems can be expressed as an elaboration of this simple setting. It is therefore useful to study this framework in great detail.

2.1.2 Two prior-to-posterior analyses

There are a number of standard families of densities that are often used to specify prior distributions over vectors of probabilities. Important such families are *conjugate* families, so those families where the prior and posterior densities in (2.1) come from the same family of distributions. These provide a powerful and consistent framework for combining prior knowledge with observational data.

A Beta prior-to-posterior analysis

We begin with the case when $\Omega = \{\omega_0, \omega_1\}$ takes just two values and we are interested in making inferences about the probabilities $p = p(\omega_0)$ and $1 - p = p(\omega_1)$. The conjugate family for a single random variable taking such a value p in the range $0 \leq p \leq 1$ is the *Beta family* Beta(α, β). Its density is given by

$$f(p|\alpha,\beta) = \frac{\Gamma(\alpha+\beta)}{\Gamma(\alpha)\Gamma(\beta)} p^{\alpha-1}(1-p)^{\beta-1} \tag{2.4}$$

where $\alpha, \beta > 0$ are hyperparameters and where $\Gamma(y) = \int_0^\infty x^{y-1}e^{-x}dx$ denotes the *Gamma* function. The Beta density (2.4) looks complicated but is in fact simply a polynomial in p. The Gamma function fraction simply ensures that f integrates to one over the unit interval. This distribution has some nice properties. For instance, the density $f(p|\alpha, \beta)$ is unimodal when $\alpha, \beta > 1$ with its mode at ${}^{\alpha-1}/_{\alpha+\beta-2}$. Its mean and variance are given by

$$\begin{aligned} \mu &= \alpha(\alpha+\beta)^{-1}, \\ \sigma^2 &= \mu(1-\mu)(\alpha+\beta+1)^{-1}, \end{aligned} \tag{2.5}$$

respectively. As the sum of the hyperparameters $\alpha + \beta$ becomes large, the conditional probability density $f(p|\alpha, \beta)$ thus concentrates its mass very close to the mean μ. In addition, the density $f(p|\alpha, \beta)$ is simply the uniform distribution when $\alpha = \beta = 1$. These two properties will be important when choosing adequate hyperparameters in a conjugate analysis: see the analysis below and the discussion in Sections 5.1 and 6.1.

Now in the virus example above assume we take a random sample of size N and we observe $N - y$ infected patients in the category ω_1 and y non-infected patients labelled as ω_0. Then when p has a Beta(α, β) prior density, using (2.1) thought of as a function of $p = p(\omega_0)$ we obtain the posterior density

$$\begin{aligned} f(p|y) &\propto f(y|p)f(p) \\ &= \binom{N}{y} p^y (1-p)^{N-y} \frac{\Gamma(\alpha+\beta)}{\Gamma(\alpha)\Gamma(\beta)} p^{\alpha-1}(1-p)^{\beta-1} \\ &\propto (1-p)^{N-y} p^y (1-p)^{N-y} p^{\alpha-1}(1-p)^{\beta-1} \\ &= p^{\alpha_+ - 1}(1-p)^{\beta_+ - 1} \end{aligned}$$

where $\alpha_+ = \alpha + y$ and $\beta_+ = \beta + (N - y)$, and the symbol $\propto$ indicates that one value is proportional to another. In particular, we can see that this posterior density is proportional to a Beta(α_+, β_+) density. But since any density must integrate to 1, this in fact implies that the posterior distribution of p given y, as in (2.1), has a Beta(α_+, β_+) density.

This observation makes this particular prior-to-posterior analysis very simple to understand and analyse. In fact, after a little rearrangement, the posterior mean μ_+ is related to both the prior mean μ and the observed sample proportion $\hat{p} = y/N$ of the feature 'not infected' via the formula

$$\mu_+ = (1 - \rho_N)\mu + \rho_N \hat{p} \tag{2.6}$$

where $0 < \rho_N = {}^{N}/_{\alpha+\beta+N} < 1$. As a consequence, this posterior mean is a *weighted average* of the prior mean—our best guess before the data—and the sample proportion. In particular, the posterior mean always lies between the prior mean and the sample proportion. The conjugate analysis above thus elegantly combines expert judgements with sampled data.

We can also see here that the larger the sample size of our data the closer is the posterior estimate to the sample proportion, so in this case maximum likelihood estimation is asymptotically the same as our Bayesian approach. However, unlike in maximum likelihood estimation, our estimate never assigns a probability of zero or one to any of the two features. We hence never preclude the possibility that for example someone not sampled might have a feature of interest whilst all the sampled units do not.

A little algebra enables us to calculate that the posterior variance is given by

$$\sigma_+^2 = \mu_+(1 - \mu_+)(\alpha + \beta + N + 1)^{-1}. \tag{2.7}$$

This variance is highest when the posterior mean is closest to $1/2$.

Whatever the value of μ_+, the posterior variance tends to zero as the sample size N gets larger and larger. We then become more and more sure that the sample proportion is close to the population mean.

This closure to sampling and the corresponding simple linear update on the hyperparameters is particularly useful. This is because from (2.1) it is easy to calculate the marginal likelihood which is intrinsic to many Bayesian methods for scoring different models giving different explanations of the underlying process. We use this explicitly in the CEG model selection methodologies in Section 6.2.

In our example, it can be shown that for $y = 0, 1, \ldots, N$, the marginal likelihood takes the form

$$f(y) = \frac{N!\Gamma(\alpha + \beta)\Gamma(\alpha + y)\Gamma(\beta + N - y)}{\Gamma(\alpha + \beta + N)y!(N - y)!\Gamma(\alpha)\Gamma(\beta)}. \tag{2.8}$$

With a uniform prior on p when $\alpha = \beta = 1$, the above reduces again to a uniform distribution on $y = 0, 1, \ldots, N$. Hence with this vague prior—with a relatively large variance—no data we might observe could be seen as unexpected compared to any other. However, with a strong prior—with a small variance—our data may well appear surprising. With many possible models to choose from we can for instance use the corresponding observed value of the marginal likelihood $f(y)$ for each model to determine which of the candidate models explain our data best: details of this are given in Section 2.1.4.

A Dirichlet prior-to-posterior analysis

We next consider the case when we plan to infer the general form of a probability mass function on $n > 2$ atoms. To make the notation more compact, we will henceforth write $p_i = p(\omega_i)$ for the probability of $\omega_i \in \Omega$ for $i = 1, 2, \ldots, n$. Suppose thus we want to make inference about a random vector of probabilities taking values $\boldsymbol{p} = (p_1, p_2, \ldots, p_n)$ on a set of n features.

Then with no additional structural domain information, a convenient family of prior distributions on the vector $\boldsymbol{p}$ is the *Dirichlet density*. The Dirichlet joint density $\mathrm{Dir}(\boldsymbol{\alpha})$ is given by

$$f(\boldsymbol{p}|\boldsymbol{\alpha}) = \frac{\Gamma(\alpha_1 + \alpha_2 + \ldots + \alpha_n)}{\Gamma(\alpha_1)\Gamma(\alpha_2)\cdots\Gamma(\alpha_n)} p_1^{\alpha_1 - 1} p_2^{\alpha_2 - 1} \cdots p_n^{\alpha_n - 1} \tag{2.9}$$

where again $p_1, p_2, \ldots, p_n \geq 0$ and $\sum_{i=1}^{n} p_i = 1$ and where $\boldsymbol{\alpha} = (\alpha_1, \alpha_2, \ldots, \alpha_n)$ is a vector of hyperparameters with positive components $\alpha_i > 0$ for $1 \leq i \leq n$ and $n \geq 2$. In particular, when $n = 2$, setting $p_1 = p$, $p_2 = 1 - p$, $\alpha_1 = \alpha$, and $\alpha_2 = \beta$ then we obtain simply the $\mathrm{Beta}(\alpha, \beta)$ distribution in (2.4). We can thus consider the Beta analysis given above as a special case of the Dirichlet prior-to-posterior analysis given below.

Throughout, we will write $\overline{\alpha} = \sum_{i=1}^{n} \alpha_i$ for the sum of components of the vector $\boldsymbol{\alpha}$. Then the mean and the variance of the vector of probabilities can then be shown to be given by the formulae

$$\mu_i = \alpha_i \overline{\alpha}^{-1} \quad \text{and} \quad \sigma_i^2 = \mu_i (1 - \mu_i)(\overline{\alpha} + 1)^{-1}. \tag{2.10}$$

In practice we need to set actual values to the vector of hyperparameters $\boldsymbol{\alpha}$ of the prior. To do this, it is custom to first elicit from a domain expert the vector $\boldsymbol{\mu} = (\mu_1, \ldots, \mu_n)$ of prior means as the best set of estimates. Only subsequently do we set the *equivalent sample size parameter* $\overline{\alpha}$. We can do this directly from the expert or through eliciting one of the variances σ_i^2—often the component $i \in \{1, \ldots, n\}$ with the largest mean [97]—and then inferring $\boldsymbol{\alpha}$ using the formulae (2.10). The larger $\overline{\alpha}$, the smaller the variance and hence the more confidence is shown in the accuracy of the prior. We discuss this further for CEGs in Sections 5.1 and 6.1.

Again by (2.1) and this time using a prior density $\mathrm{Dir}(\boldsymbol{\alpha})$, on observing $\boldsymbol{Y} = \boldsymbol{y}$ we obtain the posterior density

$$\begin{aligned} f(\boldsymbol{p}|\boldsymbol{y}) &\propto p_1^{\alpha_1 - 1} p_2^{\alpha_2 - 1} \cdots p_n^{\alpha_n - 1} p_1^{y_1} p_2^{y_2} \cdots p_n^{y_n} \\ &= p_1^{\alpha_{1+} - 1} p_2^{\alpha_{2+} - 1} \cdots p_n^{\alpha_{n+} - 1} \end{aligned}$$

where $\alpha_{i+} = \alpha_i + y_i$ for all $1 \leq i \leq n$.

This joint density is proportional to—and so equal to by the argument presented for Beta distributions—a $\mathrm{Dir}(\boldsymbol{\alpha}_+)$ density with hyperparameters $\boldsymbol{\alpha}_+ = (\alpha_{1+}, \alpha_{2+}, \ldots, \alpha_{n+})$ in the notation above. In particular, it is easy to check that the mean of this posterior density is then equal to the weighted average

$$\mu_{i+} = (1 - \rho_N)\mu_i + \rho_N \hat{p}_i$$

where $\rho_N = N(\overline{\alpha} + N)^{-1}$ and where $\hat{p}_i = {}^{y_i}/{}_N$ again denotes the proportion of the random sample with feature ω_i for $i = 1, 2, \ldots, n$. Here we can directly observe that $\rho_N = {}^1/{}_2$ if and only if $N = \overline{\alpha}$, so if and only if our sample size is precisely equal to the equivalent sample size.

Just like in the Beta analysis, as the sample size N increases—so if we observe more data—relative to $\overline{\alpha}$, we find that the posterior mean μ_{i+} gets closer and closer to the sample proportions $\hat{p}_i$ and the variance σ_i^2 converges to zero for all $1 \leq i \leq n$. As a consequence, again the maximum likelihood estimate and the Bayesian approach yield the same result and the influence of the prior becomes negligible. Because the variance of the posterior estimate converges to zero, with N large we also become very sure of this estimate.

Whenever we model probabilities as a random vector with no additional structural information then using a Dirichlet prior is especially attractive, not only because of the simple prior-to-posterior analysis we achieve as outlined above. In particular, if we choose a Dirichlet prior on probabilities $p_1, p_2, \ldots, p_n$ then we can deduce that the implied prior on any partition of the atoms is also Dirichlet, as is the distribution of any vector of conditional probabilities we might want to consider. In fact, this property is unique to this family of distributions: see for instance [37]. As a consequence, setting up priors for selection over different structural models that make consistent prior assumptions across these models is then especially easy.

With enough data we can obtain consistent estimates of any probability vector $\boldsymbol{p}$. Here of course our problem is that if N is not much larger than n then the posterior will depend heavily on the prior which itself will be based on many elicited values. Hence, when there are more than a few features and no extensive randomised data it is usually wise to incorporate various forms of *structural* prior assumptions. These will restrict the number of prior probabilities we need to estimate and so stabilise the inference. This also enables us to build into the inferential procedure a rationale that explains why we are getting the answers the methods deliver. These structural assumptions can be defined in terms of a statistical model: see Section 2.2 below.

2.1.3 Poisson–Gamma and Multinomial–Dirichlet

So far we have analysed how to update our beliefs about a system after observing a random sample. We will now present how such a sample can be obtained. In particular, we show here how units can be randomly drawn from a population using one of two distinct ways.

One standard way to sample from a population is to allocate a certain time over which to make observations and for each atom in a space

$\Omega = \{\omega_1, \omega_2, \ldots, \omega_n\}$ to record the event that someone arrived at that particular atom during that time. This regime corresponds to *Poisson* sampling n independent random variables $Y_1, Y_2, \ldots, Y_n$ with probability mass function

$$f(y_i|\phi_i) = \begin{cases} \phi_i^{y_i} \exp(-\phi_i) & \text{for } y_i = 0, 1, 2, \ldots, N \\ 0 & \text{otherwise} \end{cases} \tag{2.11}$$

where $\phi_i = y_i/N$ is the expected number of units that will be classified by atom ω_i, $i = 1, 2, \ldots, n$, and N again denotes the total number of counts. We could then perform a conjugate analysis as outlined in the previous section by placing independent Gamma(α_i, γ) prior densities on ϕ_i which have densities given by

$$f(\phi_i|\alpha_i, \gamma) = \begin{cases} \frac{\gamma^{\alpha_i}}{\Gamma(\alpha_i)} \phi_i^{\alpha_i - 1} \exp(-\gamma\phi_i) & \text{if } \phi_i > 0 \text{ for all } i = 1, \ldots, n \\ 0 & \text{otherwise} \end{cases} \tag{2.12}$$

where $\alpha_i > 0$ for $i = 1, 2, \ldots, n$ and $\gamma > 0$ are hyperparameters elicited from an expert and where Γ denotes again the Gamma function.

For this choice of priors we obtain the posterior density simply as the product

$$f(\boldsymbol{\phi}|\boldsymbol{\alpha}, \gamma) = \prod_{i=1}^{n} f(\phi_i|\alpha_i, \gamma) \tag{2.13}$$

where $\boldsymbol{\phi} = (\phi_1, \phi_2, \ldots, \phi_n)$ and $\boldsymbol{\alpha} = (\alpha_1, \alpha_2, \ldots, \alpha_n)$. It is then easy to check that the posterior joint density for the vector of expected values $\boldsymbol{\phi}$ has components which are mutually independent and all marginally Gamma distributed. Specifically, after observing a vector of data we obtain

$$f(\boldsymbol{\phi}|\boldsymbol{\alpha}, \gamma, \boldsymbol{y}) = \prod_{i=1}^{n} f(\phi_i|\alpha_{i+}, \gamma_+) \tag{2.14}$$

where $\boldsymbol{y} = (y_1, y_2, \ldots, y_n)$ are the observed counts of units observed at each of the atoms and we use the shorthand $\alpha_{i+} = \alpha_i + y_i$ and $\gamma_+ = \gamma + n$ for $i = 1, \ldots, n$.

The second method of sampling is to wait until we have observed exactly N units and then to count the number of those units that fall into the category corresponding to each atom ω_i, $i = 1, \ldots, n$. This is sometimes called *stratified* sampling. We then again perform the conjugate analysis we introduced above.

Since after observing our sample we will know the value of $N = \sum_{i=1}^{n} y_i$, a natural question to ask is whether or not there is a relationship between the inferences we draw using a prior-to-posterior analysis in these two different sampling methods.

The answer is that there is a *very* close relationship. In particular, first note that the sampling mass function of the independent Poisson variables conditioned on the event that $Y = N$ will be a Multinomial mass function as in (2.3). Then second note further that much more surprisingly the two prior-to-posterior analyses coming out of the cases described above are completely equivalent to each other. To see this define a random vector of proportions

$$\overline{\phi}^{-1}\boldsymbol{\phi} = (\overline{\phi}^{-1}\phi_1, \overline{\phi}^{-1}\phi_2, \ldots, \overline{\phi}^{-1}\phi_n) \tag{2.15}$$

where we set the random variable $\phi_i = {}^{Y_i}/_N$ to be the expected number of units in category $i = 1, \ldots, n$ and $\overline{\phi} = \sum_{i=1}^{n} \phi_i = \sum_{i=1}^{n} {}^{Y_i}/_N$ to be the expected total number of units measured. Then under the prior model above it can be shown using standard change of variable techniques that $\overline{\phi}^{-1}\boldsymbol{\phi}$ and $\overline{\phi}$ are independent where $\overline{\phi}^{-1}\boldsymbol{\phi}$ has a Dirichlet $\text{Dir}(\boldsymbol{\alpha})$ distribution and $\overline{\phi}$ has a $\text{Gamma}(\sum_{i=1}^{n} \alpha_i, \gamma)$ distribution where the values of the hyperparameters $(\boldsymbol{\alpha}, \gamma)$ of this joint density are exactly those of the original product of Gamma densities. This is also true a posteriori because of conjugacy: simply replace $(\boldsymbol{\alpha}, \gamma)$ by $(\boldsymbol{\alpha}_+, \gamma_+)$ where $\alpha_{i+} = \alpha_i + y_i$ for $i = 1, \ldots, n$.

The posterior distribution from Poisson sampling—once we learn N—is then also identical to what we would have observed had we waited for exactly N observations when we simply set $\overline{\phi}^{-1}\boldsymbol{\phi} = \boldsymbol{p}$. This property of Bayesian analyses is well known and related to uninformative stopping times. It is extremely unusual for the independence properties $\overline{\phi}^{-1}\boldsymbol{\phi} \perp\!\!\!\perp \phi$ and $\perp\!\!\!\perp_{i=1}^{k} \phi_i$ to hold simultaneously—in fact for non-degenerate random variables this is only true when $\boldsymbol{\phi}$ has the product of joint densities given in (2.13): see [33].

This Poisson–Gamma representation of the Dirichlet distribution will prove very useful when we perform Bayesian inference on CEGs in Chapter 5 and exploit the posterior densities in model selection in Chapter 6.

2.1.4 MAP model selection using Bayes Factors

Model selection is a process where we select one 'best supported model' out of a collection of k competing explanations using observed data $\boldsymbol{Y} = \boldsymbol{y}$ and given a prior distribution over the parameters associated to the model. There are many ways of doing this task, some using hypothesis testing or likelihood ratios. In this text, for convenience we perform inferences as a Bayesian and so it is natural to choose a Bayesian technique to select a best fitting model from data. Below we show how this can be done and in Section 6.2 we extend these results to CEG models.

Suppose the modeller knows that the distribution of a random vector $\boldsymbol{Y}$ must derive from one of k possible data generating mechanisms, or *models* as formally defined in Section 2.2 below. We denote these $\boldsymbol{M}_1, \boldsymbol{M}_2, \ldots, \boldsymbol{M}_k$, the prior probability of $\boldsymbol{M}_i$ generating the data being q_i for each $i = 1, \ldots, k$. Suppose further we have performed a prior-to-posterior analysis of the type illustrated in the previous section on each of these models. We show below how we can then use the joint marginal likelihood of $\boldsymbol{M}_i$ for model selection: this is the probability density $f_i(\boldsymbol{y})$ of the data $\boldsymbol{Y} = \boldsymbol{y}$ under the model $\boldsymbol{M}_i$ for $i = 1, \ldots, k$ as in (2.1).

Standard Bayesian arguments tell us that the experimenter's posterior probabilities $\boldsymbol{q}_+ = (q_{1+}, q_{2+}, \ldots, q_{k+})$ of these different models are

$$q_{i+} = \frac{q_i f_i(\boldsymbol{y})}{\sum_{j=1}^{k} q_j f_j(\boldsymbol{y})} \tag{2.16}$$

for all $i = 1, \ldots, k$. A model $\boldsymbol{M}_i$ with the largest associated value of q_{i+} is called a *Maximum A Posteriori (MAP)* model. A MAP model is thus the one which is a posteriori—taking both the observations and prior knowledge into account—the most likely to have generated the data at hand. It is therefore natural to regard this model as being our best explanation of the generating mechanism.

In order to quantify how much better than any other model a MAP model explains a given dataset, we need a score to compare two competing explanations. We thus define the *Bayes Factor (BF)* score to be the fraction of the marginal likelihood functions of two different models,

$$\mathrm{BF}(\boldsymbol{M}_i, \boldsymbol{M}_j) = \frac{f_i(\boldsymbol{y})}{f_j(\boldsymbol{y})} \tag{2.17}$$

for $i \neq j$. The smaller the Bayes Factor, the more likely model $\boldsymbol{M}_j$ than $\boldsymbol{M}_i$ has generated the data $\boldsymbol{Y} = \boldsymbol{y}$. Sometimes the BF score is expressed in terms of the log-likelihood functions, so $\mathrm{BF}(\boldsymbol{M}_i, \boldsymbol{M}_j) = \exp(\log(f_i(\boldsymbol{y})) - \log(f_j(\boldsymbol{y})))$.

From (2.16) and (2.17) we thus derive that the posterior relative probability of model $\boldsymbol{M}_i$ compared to model $\boldsymbol{M}_j$ is

$$\frac{q_{i+}}{q_{j+}} = \frac{q_i}{q_j} \cdot \frac{f_i(\boldsymbol{y})}{f_j(\boldsymbol{y})} \tag{2.18}$$

for $i \neq j$.

This posterior model score can thus be calculated by the simple formula 'prior · BF = posterior'. The expression in (2.18) is especially useful when priors are specified in terms of odds ratios [97].

Obviously to use the MAP in order to select a statistical model we need a domain expert to first specify her prior probability vector $\boldsymbol{q} = (q_1, q_2, \ldots, q_k)$ over the different candidate models. Typically each candidate model will have its own sampling distribution and will thus also need its own prior distribution over its parameters. For no particularly strong reason other than it seems to work well in practice, provided that no two statistical models are duplicated—and this can easily happen—for standard model selection techniques a common assumption is that all models are a priori equally probable.

In order to identify a MAP model we then need only use the BF score

(2.17) and take the prior into account as in (2.18). The *posterior log odds* are then equal to

$$\log(q_{i+}) - \log(q_{1+}) = \log(q_i) - \log(q_1) + \log(f_i(\boldsymbol{y})) - \log(f_1(\boldsymbol{y})). \quad (2.19)$$

We choose the model $\boldsymbol{M}_i$ maximizing this expression if $\log q_{i+} - \log q_{1+} \geq 0$ and otherwise we choose $\boldsymbol{M}_1$. This is particularly sensible when the models $\boldsymbol{M}_1, \ldots, \boldsymbol{M}_k$ are nested in the sense that each model $\boldsymbol{M}_{i-1}$ is 'simpler' than $\boldsymbol{M}_i$ for $i = 2, 3, \ldots, k$. In Section 6.2, this will mean that one model is a submodel of another.

When all models are equally probable a priori then the first term on the right hand side of (2.19) is zero. We then choose the model for which $\log(f_i(\boldsymbol{y}))$ is maximized, $i = 1, \ldots, k$. Of course there are often contextual reasons why it might be believed a priori that some models are much less likely than others. In this case the first term can and should play a dominant role and the MAP model choice will reflect this.

Although every method of model selection has its drawbacks, MAP model selection is currently very popular and will therefore be used extensively in this book. It also has some nice inferential properties. Suppose for instance the analyst must choose only one candidate model and obtains strictly positive gain only if she chooses the right one. Then if the data were generated by one of the k models she has listed it can be shown that under suitable regularity conditions and provided the data set is large enough (typically it needs to be huge even in simple models) then the MAP model will indeed choose the data-generating model. This property when applied to model selection is called *consistency*. In addition, if the data is generated by a model that is not listed in $\boldsymbol{M}_1, \ldots, \boldsymbol{M}_k$ then under the same regularity conditions the MAP model will eventually tend to the model at the shortest *Kullback-Leibler divergence* from the data generating model [9].

We will in Chapters 6 and 7 look at these methods as they apply to the fitting of BNs and CEGs to observational or experimental data. In this case we can often assume that sampled data has a prescribed form. The MAP selection based on (2.19) then often has a simple linear form which can be quickly evaluated using various tricks. Thus even when we have huge classes of candidate statistical models it is often feasible to find a good fitting explanatory model within the collection if we can set the priors on the hyperparameters in a semi-automatic way. Details will be provided in the chapters where these methods are introduced.

2.2 Statistical models and structural hypotheses

In the previous section we have only considered making inference about a probability mass function in the situation where any such mass function is a plausible one over the space of all such functions on a given set of atoms. We call the model that includes all of these probability mass functions a *saturated model.* In this saturated model we have assumed that there are no logical constraints that are embedded in the process driving the data generation and also that there are no random variables measurable with respect to the space that are strongly hypothesised to be (conditionally) independent given various features that might have occurred.

However, when formulating a model for $\boldsymbol{p}$ in practice there are usually many such structural constraints that need to be accommodated. This is helpful in the sense that it can limit the space of probability mass functions we need to consider. This makes—at least in principle—this space easier to specify and—because we need less parameters to estimate—easier to make inferences on. On the downside the specification, search and estimation of such more descriptive models needs considerable extra care.

Before we formally present the definition of a statistical model we introduce a typical albeit very simple illustration of the type of structural statistical models we focus on in this book. We will then in the following section review how such hypotheses can be embedded in a statistical model and how these allow for fast model search and inference.

2.2.1 An example of competing models

A random sample of N units is drawn from a population of individuals who have experienced two viral infections. These can be classified as either type A or type B, but not both. Let $\omega_{\mathrm{AA}}, \omega_{\mathrm{AB}}, \omega_{\mathrm{BA}}, \omega_{\mathrm{BB}}$ denote respectively the outcome that a unit presents an A and subsequently an A infection, an A and then a B virus, a type B and then a type A virus and finally a type B and then a type B again. The space of atoms to consider is hence $\Omega = \{\omega_{\mathrm{AA}}, \omega_{\mathrm{AB}}, \omega_{\mathrm{BA}}, \omega_{\mathrm{BB}}\}$. Let $y_{\mathrm{AA}}, y_{\mathrm{AB}}, y_{\mathrm{BA}}, y_{\mathrm{BB}}$ denote the respective numbers of individuals observed in each of the four categories so that $y_{\mathrm{AA}} + y_{\mathrm{AB}} + y_{\mathrm{BA}} + y_{\mathrm{BB}} = N$. Let $\boldsymbol{p} = (p_{\mathrm{AA}}, p_{\mathrm{AB}}, p_{\mathrm{BA}}, p_{\mathrm{BB}})$ denote the vector of the respective probabilities of these mutually exclusive events such that all components are probabilities and sum to one. Assume a clinician entertains four possibilities to describe this setting.

The first model $\boldsymbol{M}_0$ assumes that $(p_{\mathrm{AA}}, p_{\mathrm{AB}}, p_{\mathrm{BA}}, p_{\mathrm{BB}})$ are free to be any probability mass function at all and there is no hidden relationship in the data generating process. So $\boldsymbol{M}_0$ is the saturated model.

The second model $\boldsymbol{M}_1$ assumes that there is no dependence between the type of virus presented by a unit in her first infection and in her second: a reasonable hypothesis if the disease was fundamentally caused by the prevalence in the environment of a particular virus and everyone is equally susceptible to any such virus. If we let the random variable X_1 take the value 1 when A is present and -1 when B is present in the first infection and let X_2 take the value 1 when A is present and -1 when B is present in the second infection then this hypothesis would be equivalent to X_1 and X_2 being independent.

This is often written $X_1 \perp\!\!\!\perp X_2$. Standard probability rules tell us that this means that the components of $\boldsymbol{p}$ must satisfy the following equations

$$\begin{aligned}
p_{\mathrm{AA}} &= (p_{\mathrm{AA}} + p_{\mathrm{AB}})(p_{\mathrm{AA}} + p_{\mathrm{BA}}) \\
p_{\mathrm{AB}} &= (p_{\mathrm{AA}} + p_{\mathrm{AB}})(p_{\mathrm{AB}} + p_{\mathrm{BB}}) \\
p_{\mathrm{BA}} &= (p_{\mathrm{BA}} + p_{\mathrm{BB}})(p_{\mathrm{AA}} + p_{\mathrm{BA}}) \\
p_{\mathrm{BB}} &= (p_{\mathrm{BB}} + p_{\mathrm{BA}})(p_{\mathrm{BB}} + p_{\mathrm{AB}}).
\end{aligned}$$

A third hypothesis $\boldsymbol{M}_2$ is that any patient suffers the same type of infection the second time round irrespective of the type of infection she got first time round. This would be consistent with a conjecture that exposure to each virus made it more/less likely that the same virus was caught by that individual but otherwise there was no dependence. Here it is easy to check that if we write $X_3 = X_1 X_2$—so X_3 takes the value 1 if the types of viruses in the two infections are the same and -1 otherwise then this corresponds to the hypothesis of X_1 being independent of the new random variable X_3, so $X_1 \perp\!\!\!\perp X_3$. The probabilities are hence under this model constrained by a different set of equations, namely

$$\begin{aligned}
p_{\mathrm{AA}} &= (p_{\mathrm{AA}} + p_{\mathrm{AB}})(p_{\mathrm{AA}} + p_{\mathrm{BB}}) \\
p_{\mathrm{AB}} &= (p_{\mathrm{AA}} + p_{\mathrm{AB}})(p_{\mathrm{AB}} + p_{\mathrm{BA}}) \\
p_{\mathrm{BA}} &= (p_{\mathrm{BA}} + p_{\mathrm{BB}})(p_{\mathrm{AB}} + p_{\mathrm{BA}}) \\
p_{\mathrm{BB}} &= (p_{\mathrm{BB}} + p_{\mathrm{BA}})(p_{\mathrm{AA}} + p_{\mathrm{BB}}).
\end{aligned}$$

The final conjecture model $\boldsymbol{M}_=$ is that all states are equally probable so that $\boldsymbol{p} = (1/4, 1/4, 1/4, 1/4)$.

Now each of the $\boldsymbol{M}_1, \boldsymbol{M}_2$ and $\boldsymbol{M}_=$ defines a different structural statistical model. In addition each has its own sampling distribution with often different associated numbers of free parameters in $\boldsymbol{p}$ that need to be estimated. In particular, the saturated model $\boldsymbol{M}_0$ considered in the last section has 3 free parameters because we lose one due to the sum-to-one constraint on the atomic probabilities. It is easy to check using the equations above that the dimension of both $\boldsymbol{M}_1$ and $\boldsymbol{M}_2$ are 2. Thus most of the possible probability mass functions that might describe this example are precluded, because they do not satisfy the quadratic constraints above in order to lie in either of these models. The model $\boldsymbol{M}_=$ is a degenerate one. It is of dimension 0 and contains just the one probability mass function given above.

In the previous section we saw how to estimate $\boldsymbol{M}_0$. At a first glance, the models $\boldsymbol{M}_1$ and $\boldsymbol{M}_2$ seem more complicated to estimate since their conjugate analysis appears to break down. This is however not the case. Each of these models can be estimated using an equally elegant but nevertheless different conjugate analysis which respects the structural constraints embedded in them: see Section 2.3. Note also that each of these models can be expressed as a CEG: see Section 3.3. Each will have its own marginal likelihood. In particular, each model has an associated score which can be evaluated and the corresponding MAP estimate found as a function of the observed data. Exercise 2.1 at the end of this chapter leads through the model selection for this example.

2.2.2 The parametric statistical model

At this point it is convenient to formally define general (discrete) statistical models on which we can perform inference. We can then use this framework both to discuss the BN and also the CEG model class below.

A statistical model is simply a set of distributions over a given space. Because CEGs are discrete models over a finite space of atoms, we will provide a formal definition—which allows for an interesting geometric interpretation—of this type of model here. The CEG itself is then defined in Chapter 3. Just like in the examples discussed in Section 2.2.1, every distribution over a discrete space of atoms can be specified as a vector of the values it takes over that space. Then each entry of this vector is a value between zero and one, and the sum of all entries is equal to one. Discrete statistical models are thus sets of vectors with this property and possibly further structural constraints as above. So in formal terms, discrete models are certain subsets of a *probability simplex*

$$\Delta_{n-1} = \Big\{ \boldsymbol{p} \in \mathbb{R}^n \mid \sum_{i=1}^{n} p_i = 1, \text{ and } 0 \leq p_i \leq 1 \text{ for all } i = 1, \ldots, n \Big\} \tag{2.20}$$

where $n \in \mathbb{N}$ is the cardinality of the underlying space [29]. When distributions are assumed to be strictly positive, they lie in the *open* probability simplex $\Delta^{\circ}_{n-1} \subseteq \Delta_{n-1}$ where $p_i \in (0, 1)$ for all $i = 1, \ldots, n$. In CEG modelling, we will often assume the subset we discuss to lie in such an open simplex. This assumption enables us to avoid distracting technical issues concerning boundary cases [43].

Statistical models often depend on a set of parameters and may then be characterised as a set of distributions together with certain constraints on these parameters, for instance constraints on the mean and the variance in Gaussian models. Whenever this is the case it is vital to be able to uniquely identify each parameter in a space with one distribution in the model. We will thus work with the following definition:

Definition 2.1 (Parametric statistical model). *Let Ω always denote a finite space with $n \geq 2$ atoms $\omega \in \Omega$, and let $\Theta \subseteq \mathbb{R}^d$ denote a parameter space, $d \in \mathbb{N}$. We write $p_{\boldsymbol{\theta}} : \Omega \to (0, 1)$ for a strictly positive probability mass function on such a finite space, parametrised using $\boldsymbol{\theta} \in \Theta$. The vector of values of such a function will be denoted by the bold character $\boldsymbol{p_\theta} = \big(p_{\boldsymbol{\theta}}(\omega) \mid \omega \in \Omega\big)$ and we will call each of the components $p_{\boldsymbol{\theta}}(\omega)$, $\omega \in \Omega$, of that vector an* atomic probability. *A* discrete parametric statistical model *on Ω is then a subset*

$$\boldsymbol{M}_{\Psi} = \big\{ \boldsymbol{p_\theta} \mid \boldsymbol{\theta} \in \Theta \big\} \subseteq \Delta_{n-1} \tag{2.21}$$

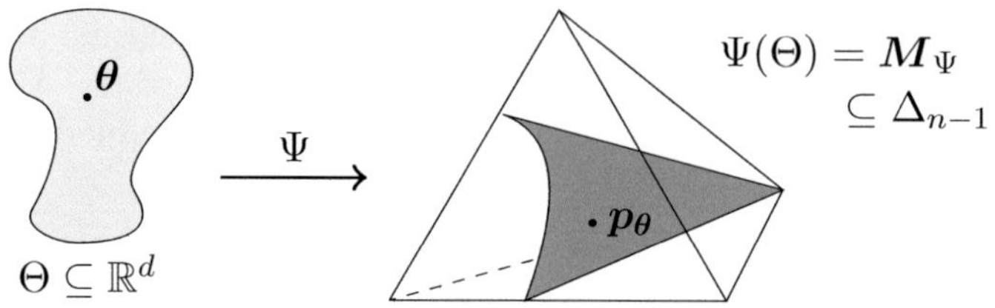

FIGURE 2.1
A discrete and parametric statistical model $\boldsymbol{M}_\Psi = \Psi(\Theta)$ depicted in dark grey as a subset of the $n-1$-dimensional probability simplex Δ_{n-1}, depicted as the tetrahedron. The model equals the image of a parametrisation $\Psi : \Theta \to \Delta_{n-1}$, $\boldsymbol{\theta} \mapsto \boldsymbol{p_\theta}$ with parameter space $\Theta \subseteq \mathbb{R}^d$ depicted in light grey.

of the $n-1$-dimensional probability simplex as in (2.20) for $n = \#\Omega$. The index Ψ in $\boldsymbol{M}_\Psi$ denotes a bijective map

$$\Psi : \ \Theta \to \boldsymbol{M}_\Psi, \quad \boldsymbol{\theta} \mapsto \boldsymbol{p_\theta} \tag{2.22}$$

which uniquely identifies a choice of parameters with a distribution in the model. Ψ is called a parametrisation *of the model $\boldsymbol{M}_\Psi = \Psi(\Theta)$ which is the image of that map.*

An illustration of the definition above is given in Figure 2.1. Naturally, there can be many different parametrisations of the same set, so often many different parametrisations of the same statistical model. We will analyse all different parametrisations giving rise to the same CEG and we will explicitly draw a number of CEG models in Section 4.2. In addition, a parametrisation of a Bayesian network—as defined in the subsequent section—is always also a parametrisation of a CEG. For both of these classes of models, the parameter space $\Theta \subseteq \mathbb{R}^d$ is itself always a product of probability simplicies, one for each marginal or conditional probability specifying the joint probability mass function: see the next section.

For graphical models such as BNs and CEGs, we will often replace the parametrisation index Ψ in $\boldsymbol{M}_\Psi$ by a graph from which we can read that parametrisation. This notation will be introduced in Chapter 3.

Note in particular that if a discrete statistical model is *saturated* as in the previous section, so if there are no constraints on the family of probability distributions in this model, then as a set of points this simply equals the full probability simplex.

The models in Section 2.2.1 were given implicitly, for instance

$$\boldsymbol{M}_2 = \{\boldsymbol{p} \in \mathbb{R}^4 \mid p_1p_4 = p_2p_3 \ \} \cap \Delta_{4-1}. \tag{2.23}$$

See Exercise 2.2 for this result. In Chapter 3, we will learn that $\boldsymbol{M}_2$ is the binary independence model on two random variables. This can be parametrised using the map $\Psi : (\theta_1, \theta_2) \mapsto (\theta_1\theta_2, \theta_1(1-\theta_2), (1-\theta_1)\theta_2, (1-\theta_1)(1-\theta_2))$

from Example 3.12. Here, the model $\boldsymbol{M}_2 = \Psi([0,1]^2)$ is the image of the unit square.

2.3 Discrete Bayesian networks

The Bayesian network (BN) was the first family of structural models to be well developed and is to this date widely used in a variety of application areas. The class of discrete BNs has a very close relationship to CEGs: it is in fact a small subclass of these models as we will show in Section 3.3. Because many of the developments enabling inference of BNs motivated our development of CEGs, it is useful to spend a little time discussing this class. There are many good books now available on Bayesian networks [22, 76, 97] and our review in this section is necessarily brief. Section 8.1 in the final chapter of this book will cover causal inference techniques for Bayesian networks which are not reviewed here.

2.3.1 Factorisations of probability mass functions

Perhaps the easiest way of thinking of a discrete Bayesian network is as a simple and convenient way of representing a factorisation of a joint probability mass function of a given vector of random variables $\boldsymbol{X} = (X_1, X_2, \ldots, X_m)$ describing a problem. Henceforth, we let $\boldsymbol{X}_A = (X_i \mid i \in A)$ denote the sub-vector of $\boldsymbol{X}$ whose indices are $A \subseteq \{1, 2, \ldots, m\}$. Assume that the dimension of the state space $\mathbb{X}_i$ of X_i is n_i for all $i = 1, 2, \ldots, m$, so that the dimension of the product space $\mathbb{X} = \mathbb{X}_1 \times \mathbb{X}_2 \times \ldots \times \mathbb{X}_m$ equals $n = \prod_i n_i$: often a huge number. In this development n will thus denote the cardinality of the space Ω of possible outcomes $\boldsymbol{X} = \boldsymbol{x}$, and m will denote the number of random variables in a model.

Introducing variables one at a time and in order, from the usual rules of probability the joint mass function or density of $\boldsymbol{X}$ can be written as the product of conditional mass functions

$$p(\boldsymbol{x}) = p_1(x_1)p_2(x_2|x_1)p_3(x_3|x_1,x_2)\cdots p_m(x_m|x_1,x_2,\ldots,x_{m-1}) \qquad (2.24)$$

for all $\boldsymbol{x} \in \mathbb{X}$. Here, $p_1(x_1)$ denotes the probability mass function of x_1 whilst $p_i(x_i|x_1, x_2, \ldots, x_{i-1})$ denotes the probability mass function of x_i conditional on the values of the components of $\boldsymbol{x}$ listed before it. To ease the explanation below, we will henceforth assume that all the probabilities in this product space are strictly positive. When thinking of these conditional probabilities as parameters, (2.24) specifies a parametric statistical model as in Definition 2.1 with $\Omega = \mathbb{X}$: see also Section 3.3.

When all the components of $\boldsymbol{X}$ are independent then the probability mass function (2.24) takes the form of a product $p(\boldsymbol{x}) = \prod_{i=1}^{m} p_i(x_i)$ of marginal probabilities. However, in most interesting models variables are typically not all independent of each other. In these models for any given $i = 1, 2, \ldots, m$, many of the functions $p_i(x_i|x_1, x_2, \ldots, x_{i-1})$ in (2.24) will be an explicit function *only* of components of $\boldsymbol{X}$ whose indices lie in some proper subset $Q_i \subsetneq \{1, 2, \ldots, i-1\}$ of the indices above. Thus suppose

$$p_i(x_i|\boldsymbol{x}_{\{1,\ldots,i-1\}}) = p_i(x_i|\boldsymbol{x}_{Q_i}) \tag{2.25}$$

for $i = 2, \ldots, n$. Then writing the index set $\{1, 2, \ldots, i-1\} = Q_i \cup R_i$ as the disjoint union of the *parent set* Q_i above with the *remainder set* $R_i = \{1, 2, \ldots, i-1\} \setminus Q_i$, we find that

$$p_i(x_i|\boldsymbol{x}_{R_i}, \boldsymbol{x}_{Q_i}) = p_i(x_i|\boldsymbol{x}_{Q_i}) \tag{2.26}$$

where in general we allow both Q_i and R_i to be empty, $i = 1, 2, \ldots, m$.

What we have asserted here is that many of the conditional probabilities $\{p_i(x_i|x_1, \ldots, x_{i-1}) \mid i = 1, 2, \ldots, m\}$ that determine the probability mass function $p(\boldsymbol{x})$ on the atoms of the space are set equal to one another. An equivalent way of stating this is to assert that for each possible value of $\boldsymbol{x}_{Q_i} \in \mathbb{X}_{Q_i}$ and for all possible values of $x_i \in \mathbb{X}_i$, the conditional probabilities

$$p_i(x_i|\boldsymbol{x}_{R_i}, \boldsymbol{x}_{Q_i}) = p_i(x_i|\boldsymbol{x}'_{R_i}, \boldsymbol{x}_{Q_i}) \tag{2.27}$$

are equal for any two possible configurations $\boldsymbol{x}_{R_i}$ and $\boldsymbol{x}'_{R_i}$ of the remainder vector, $i = 1, 2, \ldots, m$. It is the exact analogy of this type of property—where various vectors of probabilities are asserted to be equal—which we use to define the staged tree and CEG in the next chapter.

By substituting (2.25) into (2.24), we can now obtain a new factorisation of the probability mass function

$$p(\boldsymbol{x}) = p_1(x_1) \prod_{i=2}^{m} p_i(x_i|\boldsymbol{x}_{Q_i}) \tag{2.28}$$

so that the corresponding statistical model is parametrised by the set of conditional probabilities $\{p_i(x_i|\boldsymbol{x}_{Q_i}) \mid i = 1, 2, \ldots, m\}$. These parameters will be denoted $\theta(x_i, \boldsymbol{x}_{Q_i})$ in Section 3.3.

Taking account of the fact that probabilities must sum to one, the identification in (2.27) gives us a submodel—a family of distributions as in (2.28)—of the saturated model with $\sum_{i=1}^{m}(n_i - 1)\prod_{j \in Q_i} n_j$ rather than $\prod_{j=1}^{m} n_j - 1$ free parameters: so often a much lower dimensional model.

Directly from the definition of conditional independence, we can see that (2.27) is equivalent to a specification of the model via a set of $m-1$ irrelevance statements over the components of $\boldsymbol{X}$:

$$X_i \perp\!\!\!\perp \boldsymbol{X}_{R_i} \mid \boldsymbol{X}_{Q_i} \qquad \text{for all } i = 2, \ldots, m \tag{2.29}$$

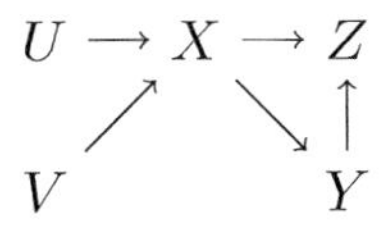

FIGURE 2.2
A DAG for Example 2.3.

where we use the convention that $\boldsymbol{X} \perp\!\!\!\perp \boldsymbol{Y}|\boldsymbol{Z}$ reads that $\boldsymbol{X}$ is independent of $\boldsymbol{Y}$ conditional on $\boldsymbol{Z}$. In other words, once we know $\boldsymbol{Z}$, the random variable $\boldsymbol{Y}$ provides no further information about $\boldsymbol{X}$. The central insight in Bayesian networks is that this collection of conditional independence statements can be expressed unambiguously and in a rather obvious way by a *directed acyclic graph (DAG)* where vertices are random variables and the absence of edges indicates conditional independence. This in particular then gives a graphical representation of the factorisation (2.28) that we use to represent a BN model. For this reason, a BN is also called a *graphical model.*

Definition 2.2 (Bayesian network). *A* Bayesian network (BN) *on the random variables $X_1, X_2, \ldots, X_m$ is a set of the $m-1$ conditional independence statements (2.29)—or equivalently a family of probability distributions with the factorisation (2.28)—together with a DAG $\mathcal{G} = (V, E)$. The set of vertices V of the DAG is given by $\{X_1, X_2, \ldots, X_m\}$ and a directed edge from X_i into X_j is in the edge set E of $\mathcal{G}$ if and only if X_i is a component of the vector $\boldsymbol{X}_{Q_j}$ where Q_j is the parent set of X_j for all $1 \leq i, j \leq m$.*

A DAG $\mathcal{G}$ is said to be valid *if the decision maker believes the conditional independence statements associated with its BN.*

A Bayesian network model or BN model *is the family of probability distributions which fulfill the conditional independence assumptions coded in a BN. A BN model is said to be* represented by *the DAG of the BN.*

The graph constructed in Definition 2.2 is automatically acyclic because a variable vertex can only lead into another vertex with a higher index. A typical example of such a BN is given below.

Example 2.3 (A first DAG). *Consider five random variables taken in the order (U, V, X, Y, Z). Then the DAG in Figure 2.2 corresponds to the factorisation of the joint probability mass function of U, V, X, Y, Z given by*

$$p(u, v, x, y, z) = p(u)p(v)p(x|u, v)p(y|x)p(z|x, y)$$

for all states u, v, x, y, z the random variables above can take.

The alternative ordering (V, U, X, Y, Z) would give the same graph and the same factorisation of $p(u, v, x, y, z)$. Although we cannot necessarily read a unique ordering from a DAG of a BN it can be shown that all such orderings must correspond to the same factorisation and so, when all the sample spaces are known, in particular to the same statistical model [63].

There has been an interesting development over recent years partly induced by the need to build larger and larger conditional independence models as above. No matter how large the sample space of such a BN is, there is a limit to the number of conditional probabilities that can be elicited or reliably estimated. It had long been noted however that many more of the conditional probability vectors—and not just those associated with particular configurations of parent vertices in the BN as in (2.27)—could be conjectured to be the same. If this is allowed then the corresponding statistical model immediately becomes more parsimonious and easy to fit. We call this type of model a *context-specific* Bayesian network [12, 62]. The use of such symmetries have greatly expanded the potential use of the BN machinery.

However, the graphical description given above of a model class then becomes incomplete: it can only accommodate some of the features in the factorisation of the probability mass function. So then the DAG representation becomes less useful. The CEG on the other hand includes all such context-specific information in a BN in its own graph and colouring: see Section 3.3. In this way we can resurrect the use of graphs to better understand these wider classes of models.

2.3.2 The d-separation theorem

Graphs are useful because they allow us to visually appreciate the conditional independence information implicit in a given factorisation. In particular, this type of information can be expressed in natural language and so is easier to elicit. The *d-separation theorem* first developed by [40, 109] and then re-expressed in the form given here by [63, 65] is a powerful result that enables us to analyse logical implications of a specified BN from its graph alone.

Before we can articulate this result we need a few terms from graph theory. These notions and in particular the notation we use in this section will not be central to the remainder of the book but are repeated here exclusively to be able to unambiguously state the result below. In Section 4.1 we can then present a theorem for CEGs which is analogous to d-separation in BNs.

Let $\mathcal{G} = (V, E)$ be a DAG and let its vertex set V include the random variables X, Y, Z and the vector $\boldsymbol{X}$. A vertex X is a *parent* of a vertex Y, and Y is a *child* of X in the directed graph $\mathcal{G}$ if and only if there is a directed edge $X \to Y$ from X to Y in $\mathcal{G}$. Similarly, we say Z is an *ancestor* of Y in the directed graph $\mathcal{G}$ if $Z = Y$ or if there is a directed path in $\mathcal{G}$ from Z to Y. This terminology can also be made to apply to all subsets of the vertex set $V = V(\mathcal{G})$. In particular, the *ancestral set* $A(\boldsymbol{X}) \subseteq V(\mathcal{G})$ of vertices $\boldsymbol{X}$ is the set of all the vertices that are ancestors of a vertex in $\boldsymbol{X}$. The *ancestral graph* $\mathcal{G}(A(\boldsymbol{X}))$ then has vertex set $V(\mathcal{G}(A(\boldsymbol{X})) = A(\boldsymbol{X})$ and edge set $E(\mathcal{G}(A(\boldsymbol{X})) = \{X \to Y \in E(\mathcal{G}) \mid X, Y \in A(\boldsymbol{X})\}$. Thus an ancestral graph $\mathcal{G}(A(\boldsymbol{X}))$ is the subgraph of $\mathcal{G}$ generated by the ancestral set $A(\boldsymbol{X})$ of $\boldsymbol{X}$.

A graph is said to be *mixed* if some of its edges are directed and some are undirected. The *moralised graph* $\mathcal{G}^{\mathrm{m}}$ of a directed graph $\mathcal{G}$ has the same vertex set and set of directed edges as $\mathcal{G}$ but has an undirected edge between any two vertices for which there is no directed edge between them in $\mathcal{G}$ but both are parents of the same child in $V(\mathcal{G})$. Thus, continuing the hereditary metaphor, all unjoined parents of each child are 'married' together in this operation. If $\mathcal{G}^{\mathrm{m}} = \mathcal{G}$—so that all two parents of the same child are joined by a directed edge for all children in $V(\mathcal{G})$—then $\mathcal{G}$ is said to be *decomposable*. The *skeleton* $\mathcal{S}(\mathcal{H})$ of a mixed graph $\mathcal{H}$ is one with the same vertex set as $\mathcal{H}$ and an undirected edge between X and Y if and only if there is a directed or undirected edge between X and Y in $\mathcal{H}$. Thus to produce the skeleton $\mathcal{S}(\mathcal{H})$ of a mixed graph $\mathcal{H}$ we simply replace all directed edges in $\mathcal{H}$ by undirected ones. Finally, for any three disjoint subsets $A, B, C \subseteq \{1, 2, \ldots, m\}$ and $\boldsymbol{X}_A, \boldsymbol{X}_B, \boldsymbol{X}_C$ the corresponding sets of the vertices $V(\mathcal{S})$ of an undirected graph $\mathcal{S}$, we say that $\boldsymbol{X}_B$ does *separate* $\boldsymbol{X}_C$ from $\boldsymbol{X}_A$ in $\mathcal{S}$ if and only if any path from any vertex X_a with $a \in A$ to any vertex X_c with $c \in C$ passes through a vertex X_b with $b \in B$. We then have the following.

Proposition 2.4 (d-separation). *Let A, B, C be any three disjoint subsets of an index set $\{1, 2, \ldots, m\}$ and let $\mathcal{G}$ be a valid DAG with vertices $V(\mathcal{G}) = \{X_1, X_2, \ldots, X_m\}$. If $\boldsymbol{X}_B$ separates $\boldsymbol{X}_C$ from $\boldsymbol{X}_A$ in the skeleton of the moralised graph $\mathcal{G}^{\mathrm{m}}(A(\boldsymbol{X}_{A \cup B \cup C}))$ of the ancestral graph of $\boldsymbol{X}_{A \cup B \cup C}$ then this is equivalent to $\boldsymbol{X}_C \perp\!\!\!\perp \boldsymbol{X}_A \mid \boldsymbol{X}_B$.*

The result above enables the modeller to directly infer conditional independence statements from the DAG representing a BN model, without the need to specify a probability mass function or indeed infer values of such a function. This makes these types of graphical models extremely powerful: a property that extends to CEGs. In particular, throughout Chapter 4 we demonstrate how various implicit conditional independences can be read from a CEG that give us insight over some of the global properties of its associated statistical model.

2.3.3 DAGs coding the same distributional assumptions

Two BN models as in Definition 2.2 can make the same distributional assumptions but have different DAG representations—just like two statistical models in Definition 2.1 can have different parametrisations. DAGs representing the same model will here be called *statistically* equivalent. For example, the three DAGs below are different and each embodies a different factorisation of the associated probability mass function but all embody just the two conditional independence statements $X \perp\!\!\!\perp Y \mid Z$ and $Y \perp\!\!\!\perp X \mid Z$. The family of possible joint models consistent with each are identical:

$$\mathcal{G}_1 : \ X \to Z \to Y \qquad \mathcal{G}_2 : \ X \to Z \to Y \qquad \mathcal{G}_3 : \ X \to Z \to Y$$

It was proved by [109] that two BNs are associated with the same statistical model if and only if their DAGs share the same *pattern*. The pattern $\mathcal{P} = \mathcal{P}(\mathcal{G})$

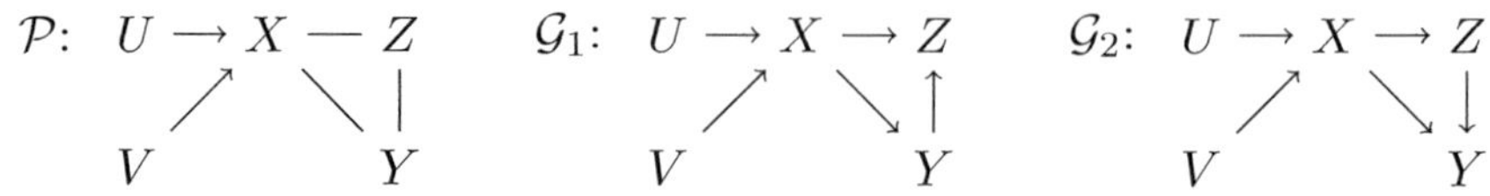

FIGURE 2.3
A pattern and two equivalent DAGs with this pattern.

of a DAG $\mathcal{G}$ is a mixed graph with same vertex set $V(\mathcal{P}) = V(\mathcal{G})$ and a directed edge $X \to Y \in E(\mathcal{G})$ from X to Y replaced by an undirected edge X—$Y \in E(\mathcal{P})$ if and only if there exists no other parent of Y which is not connected to or from X by an edge. In simple words, two DAGs represent the same statistical model if they have the same skeleton and the same configurations of unmarried parents: see the previous section for this terminology. A slightly bigger example is given in Figure 2.3 which shows the pattern of the DAG from Figure 2.2 together with the only other two DAGs that share this pattern. The corresponding statistical model can thus be represented by precisely these three DAGs.

Thanks to this characterisation, when searching across a model space of BNs we can ensure that any two equivalent model representations are scored the same: see Section 2.1 and our development in Chapter 5. The easiest way to do this is to restrict the priors we use to Dirichlet distributions on all edge probabilities associated with each configuration of values of a parent and to choose the hyperparameters appropriately.

We will see in Section 4.2 that in a model search over classes of Chain Event Graphs two distinct such coloured graphs can also represent the same statistical model: so similar inferential care is required. However, the equivalence classes of CEGs associated with a statistical model cannot be easily represented through a modified graph such as the pattern introduced above. They can instead in a different but equally elegant way be indexed by a generating function of polynomial form [47].

2.3.4 Estimating probabilities in a BN

Very often we need to estimate a collection of conditional probabilities $\{p_i(x_i|\boldsymbol{x}_{Q_i}) \mid i = 1, 2 \ldots, m\}$ that define a BN model represented by a DAG $\mathcal{G}$ as in (2.28). A natural question to ask is whether it is possible to do this in an elegant way just as for the saturated model of Section 2.1. We show below that the answer to this question is yes and provide the supporting theory.

In a BN model we will need to place a prior on each of the probability vectors of each X_i given all the possible values $\boldsymbol{x}_{Q_i}$ of its parents. Just like in Section 2.1.1, under multinomial $\text{Multi}(N, p(\boldsymbol{x}))$ sampling of a random variable $\boldsymbol{Y}$ conditional on the set $\{p(\boldsymbol{x}) \mid \boldsymbol{x} \in \mathbb{X}\}$ of atomic probabilities, the likelihood function will be

$$f(\boldsymbol{y}|p(\boldsymbol{x})) = \frac{N!}{y_1!y_2!\dots y_n!} \prod_{\boldsymbol{x}\in\mathbb{X}} p(\boldsymbol{x})^{y(\boldsymbol{x})} \tag{2.30}$$

where $p(\boldsymbol{x}) \geq 0$ for $\boldsymbol{x} \in \mathbb{X}$ and $\sum_{\boldsymbol{x}\in\mathbb{X}} p(\boldsymbol{x}) = 1$ and $\boldsymbol{y} = \boldsymbol{y}(\boldsymbol{x})$ is the observed realisation of $\boldsymbol{Y}$. Of course here the number n of atoms in $\mathbb{X}$ is typically huge. As a consequence, most of the counts $y(\boldsymbol{x})$ of units in the sample which actually take the values $\boldsymbol{x}$ are usually small and are often equal to one. However, if $\mathcal{G}$ is valid then we can plug in the factorisation (2.28) and (2.30) simplifies to

$$f(\boldsymbol{y}|p(\boldsymbol{x})) = \frac{N!}{y_1!y_2!\dots y_n!} \prod_{\boldsymbol{x}\in\mathbb{X}} \left(p_1(x_1)\prod_{i=2}^{m} p_i(x_i|\boldsymbol{x}_{Q_i})\right)^{y(\boldsymbol{x})}. \tag{2.31}$$

We thus end up with what is called a *separable likelihood.* This form of a likelihood is especially useful because it enables us to split up information about the data into a product of functions where each component in that product informs exactly one of the components of a given parameter vector: here the probability of a particular component random variable given a particular configuration of its parents. This makes estimation particularly simple.

The likelihood function above can then be equivalently written as

$$f(\boldsymbol{y}|p(\boldsymbol{x})) \propto f_1(\boldsymbol{y}|p_1(x_1)) \prod_{i=2}^{m} \prod_{\boldsymbol{x}_{Q_i}\in\mathbb{X}_{Q_i}} f_i(\boldsymbol{y}|p_i(x_i|\boldsymbol{x}_{Q_i})) \tag{2.32}$$

where

$$f_1(\boldsymbol{y}|p_1(x_1)) = \prod_{\boldsymbol{x}_1\in\mathbb{X}_1} p_1(x_1)^{y(x_1)}$$
$$f_i(\boldsymbol{y}|p_i(x_i|\boldsymbol{x}_{Q_i})) = \prod_{\boldsymbol{x}_i\in\mathbb{X}_i} p_i(x_i|\boldsymbol{x}_{Q_i})^{y(\boldsymbol{x}_i|\boldsymbol{x}_{Q_i})}$$

and where $y(x_1)$ denotes the number of units in the sample for whom $X_1 = x_1$ and $y(\boldsymbol{x}_i|\boldsymbol{x}_{Q_i})$ denotes the number of units whose parent configuration is $\boldsymbol{x}_{Q_i}$ and for which $X_i = x_i$ for $i = 2, 3, \dots, m$.

Now analogous to the more general Bayesian updating in Section 2.1.1, suppose we put Dirichlet $\mathrm{Dir}(\boldsymbol{\alpha}_1)$ prior density $f(\boldsymbol{p}_1|\boldsymbol{\alpha}_1)$ on the marginal probabilities $\{p_1(x_1) \mid x_1 \in \mathbb{X}_1\}$ where $\boldsymbol{\alpha}_1 = (\alpha_{11}, \alpha_{12}, \dots, \alpha_{1n_1})$. Then for each $i = 2, 3, \dots, m$ and each possible configuration $\boldsymbol{x}_{Q_i} \in \mathbb{X}_{Q_i}$ we put again Dirichlet $\mathrm{Dir}(\boldsymbol{\alpha}_i(\boldsymbol{x}_{Q_i}))$ prior density $f_i(\boldsymbol{p}_i|\boldsymbol{\alpha}_i(\boldsymbol{x}_{Q_i}))$ on the conditional probability vectors $\{p_i(x_i|\boldsymbol{x}_{Q_i}) \mid x_i \in \mathbb{X}_i\}$ where $\boldsymbol{\alpha}_i(\boldsymbol{x}_{Q_i}) = (\alpha_{i1}(\boldsymbol{x}_{Q_i}), \alpha_{i2}(\boldsymbol{x}_{Q_i}), \dots, \alpha_{in_i}(\boldsymbol{x}_{Q_i}))$. Often all of these entries are set to unity, in practice giving a uniform prior distribution over this probability simplex—and obviously we will do better if we use prior informed expert judgements to choose these [97]. Assuming that all the conditional probability vectors are a priori mutually independent of each other—this is the so called *local and global independence assumption*—the joint prior over these probabilities is simply the product of these densities.

In this setting, using Bayes' Rule on the vectors of joint and conditional probabilities we immediately see that—since the likelihood is also a product of functions in these conditional probability vectors and the posterior density is a product of the likelihood and prior as in (2.1)—the posterior densities must take the form of a product of Dirichlet densities

$$f(\boldsymbol{p}|\boldsymbol{\alpha}_1, \boldsymbol{y}(\boldsymbol{x})) = f_1(\boldsymbol{p}_1|\boldsymbol{\alpha}_{1+}) \prod_{i=2}^{n} \prod_{\boldsymbol{x}_{Qi} \in \mathbb{X}_{Q_i}} f_i(\boldsymbol{p}_i|\boldsymbol{\alpha}_{i+}(\boldsymbol{x}_{Q_i})) \tag{2.33}$$

where we again denote $\boldsymbol{\alpha}_{1+} = \boldsymbol{\alpha}_1 + y(\boldsymbol{x}_1)$ and $\boldsymbol{\alpha}_{i+}(\boldsymbol{x}_{Q_i}) = \boldsymbol{\alpha}_i(\boldsymbol{x}_{Q_i}) + y(\boldsymbol{x}_i|\boldsymbol{x}_{Q_i})$ for each $\boldsymbol{x}_{Q_i} \in \mathbb{X}_{Q_i}$ and $i = 2, 3, \ldots, m$.

As a consequence, just like in the saturated case in Section 2.1.2, with this product Dirichlet prior we have a very elegant and transparent updating formula also for BN models. In particular, random sampling has not destroyed the conditional prior mutual independence relationship between the conditional probabilities in the process: local and global independence is also a property of this posterior distribution.

We can also calculate the marginal likelihood of this BN. It is easy to show that this is simply the product of the marginal likelihood obtained from a Bayesian analysis of the separate conditional probability components. This means that the log marginal likelihood score is linear in the log of the marginal likelihood score of the component models. The closed form of the score and this linearity makes the space of all such models amenable to some very fast search algorithms across the space of all BNs on the components of $\boldsymbol{X}$ [22].

Despite the advantages that we have here, when problems get big the number of parameters in these models is still very large and so the data informing the different parts of the model can be very sparse. This can often be overcome by assuming further context-specific information that hypothesises that more of the probability vectors appearing in these factorisations can be associated with each other. Assuming this in fact makes no substantive change in this conjugate analysis and similar formulae then apply. Since this analysis is just a special case of results we report about the CEG we will defer discussion of this issue to Chapter 5. We show there that even in problems with very asymmetric state spaces exactly analogous conjugate analyses can apply.

2.3.5 Propagating probabilities in a BN

One of the first issues in statistical inference that could be addressed through BNs was to now have available a framework that could be used to quickly and efficiently calculate various conditional probabilities for a given probability mass function p. We have seen in the previous section that if a DAG is valid then its associated statistical model is parametrised by the set of conditional probabilities $\{p_i(x_i|\boldsymbol{x}_{Q_i}) \mid i = 1, 2 \ldots, m\}$. Assume that all these conditional probabilities have been specified and do not need to be estimated. Using a naïve approach, we can of course simply multiply these together using the formula (2.28) to obtain the full joint distribution of $p(\boldsymbol{x})$. If we then want to find the distribution of $\boldsymbol{x}$ given some observation $\boldsymbol{y}(\boldsymbol{x})$ of the system then we can use Bayes' Rule to calculate the probabilities $p(\boldsymbol{x}|\boldsymbol{y})$ of interest.

The sort of scenario we have in mind here concerns for instance the health indicators of a medical patient—described by a very long vector $\boldsymbol{X}$ of variables used for the description of potential diseases and symptoms they might present and report. Suppose we have available to us a very large BN that applies to a population of patients to which the patient before us belongs and suppose that the population's conditional probabilities are all known. We now observe a subset $\boldsymbol{y}(\boldsymbol{x})$ of the possible symptoms that this patient exhibits. On this new evidence we want to revise our probabilities about this particular patient and want to calculate the new probabilities that pertain to the marginal distribution of certain components of $\boldsymbol{x}$, especially those associated with a disease.

Using a brute force method to solve this type of problem requires many computations. Furthermore, if the vector $\boldsymbol{x}$ is very long then the sample space could be huge, so even storing the numbers in $p(\boldsymbol{x})$ could be prohibitive. The key question is: *Is there a way to use the BN to calculate the conditional distributions of certain posterior margins of $\boldsymbol{x}$ more quickly?* And can we write an algorithm that will work generically to calculate the conditional distribution of any function of target variables given any vector $\boldsymbol{y}(\boldsymbol{x})$ of observations that might be presented?

The most general answer to this question is *no*. However, if we suitably restrict the question to apply only to certain BNs and observations of certain types and target variables then there are various simple algorithms which can do precisely this. These algorithms are then called *propagation algorithms*.

The most used propagation algorithm for BNs is called the *junction tree algorithm* [56]. In its vanilla form it assumes the following

1. That a single given BN $\mathcal{G}$ describes the process and that this BN is decomposable: see Section 2.3.2.
2. That the event that we condition on is the value $\boldsymbol{X}_A = \boldsymbol{x}_A$ for an arbitrary possible subset $A \subseteq \{1, \ldots, m\}$ of the components of $\boldsymbol{X}$.
3. That the *target probability* is the probability of some arbitrary component X_i of $\boldsymbol{X}$ where $i \notin A$.

The standard junction tree algorithm first transforms $\mathcal{G}$ into a junction tree $\mathcal{J} = \mathcal{J}(\mathcal{G})$ which is a DAG whose vertices are given by the cliques of $\mathcal{G}$ and edges given by the separators defined below. The algorithm then transforms the original conditional probability tables $\{p_i(x_i|\boldsymbol{x}_{Q_i}) \mid i = 1, 2 \ldots, m\}$ into marginal distributions over selected subvectors of $\boldsymbol{X}$ which are associated to cliques. In this development, *cliques* $C_j \subseteq \{1, \ldots, m\}$ are maximally connected subsets of vertices of $\mathcal{G}$, $j = 1, 2, \ldots, k$. These are such that their union covers all indices $\bigcup_{j=1}^{k} C_j = \{1, 2, \ldots, m\}$ so in particular each component of $\boldsymbol{x}$ is part of at least one such clique. Since $\mathcal{G}$ is decomposable, it is possible to identify a collection of $k-1$ subsets $B_j \subsetneq C_j$, where $j = 2, 3, \ldots, k$, for which each such *separator* B_j is an intersection of cliques so that we can write the associated probability mass function as

$$p(\boldsymbol{x}) = \frac{\prod_{j=1}^{m} p_{C_j}(\boldsymbol{x}_{C_j})}{\prod_{j=2}^{m} p_{B_j}(\boldsymbol{x}_{B_j})} \tag{2.34}$$

for all atoms $\boldsymbol{x} \in \mathbb{X}$. As an aside note that the parameters in (2.34) will be marginally denoted by θs in this text—they are precisely the ones we use to specify a BN as a parametric statistical model in Section 3.3.

It can be shown that such a junction-tree representation of the original BN remains valid after conditioning on events of the form $\boldsymbol{y}(\boldsymbol{x})$ given above. Hence,

$$p(\boldsymbol{x}|\boldsymbol{y}(\boldsymbol{x})) = \frac{\prod_{j=1}^{m} p_{C_j}(\boldsymbol{x}_{C_j}|\boldsymbol{y}(\boldsymbol{x}))}{\prod_{j=2}^{m} p_{B_j}(\boldsymbol{x}_{B_j}|\boldsymbol{y}(\boldsymbol{x}))}. \tag{2.35}$$

Determining a formula that calculates the conditional probabilities $p_{C_j}(\boldsymbol{x}_{C_j}|\boldsymbol{y}(\boldsymbol{x}))$ for $j = 1, 2, \ldots, k$ above as a function of marginal probabilities over cliques $\{p_{C_j}(\boldsymbol{x}_{C_j}) \mid j = 1, 2, \ldots, k\}$, in particular will give us the posterior marginal distributions of each component of $\boldsymbol{X}$ that we are interested in. This is achieved by a sequence of operations called *collect* and *distribute*, performed as a map from the current clique margin to one sharing an index in a separator. There are many such methods now coded in various software such as `R` [82]. There are also other collections of techniques—like *lazy propagation algorithms* [57]—with different types of conditioning events and target probabilities: but all of which frame such needs in terms of collections of values of the random subvectors defining the problem a priori.

The methods proposed above, at least in their vanilla form, work only for certain types of information—here the intersection of functions of the cliques of the triangulated BN—and on particular types of target variables—here functions of variables in a single clique. For excellent reviews and description of some of these algorithms see for example [22] or [57]. In problems where the components of the random vector $\boldsymbol{X}$ naturally describe the process of interest, questions of a level of generality which can be answered by these propagation algorithms—so those that meet the three assumptions above—are natural questions to be interested in. These will provide at least a subset of the questions a practitioner might want answered. However, in many real scenarios they are still quite restrictive. For example, propagation does not directly answer questions concerning observations where symptoms are only partially observed—for example when we learn that one of two component binary variables takes the value 1 but we do not observe which, or the target variable is a random variable which is the sum of disparate components of $\boldsymbol{X}$.

The obvious question here is whether we can devise similar algorithms for a structural model defined in terms of a CEG. The answer is that *yes*, we can. In fact, we can use the graph of the CEG itself as a framework for the local information passing simply forgetting its colours—without the need to construct a new graph like the junction tree. Of course for problems represented by CEGs the sets of target probabilities and conditioning events the generic algorithm serves tend to be different from those served by a BN. This is because a tree model is often built to answer different questions. In particular, the algorithm does not specify these in terms of a product space of a predetermined set of variables. However, there are many similarities. This development is outlined in Section 5.2.

2.4 Concluding remarks

We have seen in the final section of this chapter that standard methods of Bayesian inference used in conjunction with structural information—often coded in terms of a DAG—provide a powerful framework for expressing multivariate relationships, for making inferences within a model and also for selecting good explanatory models. However, many standard graphical frameworks such as BNs, chain graphs, ancestral graphs and others start with the hypothesis that we are first given a set of random variables and that the structural relationships we embed in the model are about whether or not one vector of these variables is relevant to the forecasts of another. Often however, as we have illustrated in Chapter 1 of this book, we think in stories about how an event might unfold. Although we can talk about dependence in this context, the natural framework to express our beliefs is then not a DAG over a preassigned set of variables but an event tree which does not require us to specify these variables.

So in a problem that is naturally and fully expressed in terms of a vector of component random variables and their conditional independence relationships, the methods supporting these—like the ones illustrated throughout Section 2.3—should be used. However, when this is not the case we should check whether the framework offered by the Chain Event Graph—as presented through Chapters 4 to 6—might be better used for elicitation, inference and model discovery.

In Chapter 3 we next show formally how we can use these event trees as an inferential framework just like a BN but which can process and transmit information that is embedded in stories rather than collections of dependence hypotheses. This book then develops powerful analogues of all the techniques discussed above that are able to address a much wider class of problems than a BN and work almost as simply and transparently.

2.5 Exercises

Exercise 2.1. *For the saturated model* $\boldsymbol{M}_0$ *in Section 2.2.1 use a Dirichlet* $\text{Dir}(1,1,1,1)$ *prior and find the marginal likelihood explicitly given the cell counts* $(y_{\text{AA}}, y_{\text{AB}}, y_{\text{BA}}, y_{\text{BB}})$.

For model selection it is often argued that the prior distributions over the shared hyperparameters—here various marginal probabilities—should be set equal to one another if possible. Show that in this case we should set the distribution of the prior probabilities $P(X_1 = 1)$, $P(X_2 = 1)$ *within* $\boldsymbol{M}_1$ *each as* $\text{Beta}(2,2)$ *and* $P(X_1 = 1)$, $P(X_3 = 1)$ *each as* $\text{Beta}(2,2)$ *in* $\boldsymbol{M}_2$.

Assuming these probabilities are each independent a priori within each model $\boldsymbol{M}_1$ *and* $\boldsymbol{M}_2$, *calculate the regions* $(y_{\text{AA}}, y_{\text{AB}}, y_{\text{BA}}, y_{\text{BB}})$ *of the highest scoring model in each case.*

Exercise 2.2. *In Section 2.2.1 show that* $\boldsymbol{M}_1$ *assumes only that* $p_{\text{AA}}p_{\text{BB}} = p_{\text{AB}}p_{\text{BA}}$ *whilst model* $\boldsymbol{M}_2$ *assumes only that* $p_{\text{AA}}p_{\text{BA}} = p_{\text{AB}}p_{\text{BB}}$. *Write down the single corresponding constraint associated with the model* $\boldsymbol{M}_3$ *that demands that* $X_2 \perp\!\!\!\perp X_3$. *Hence show that* $\boldsymbol{M}_= = \boldsymbol{M}_1 \cap \boldsymbol{M}_2 \cap \boldsymbol{M}_3$.

Exercise 2.3. *Plot the solution sets of the model-defining equations specified in Exercises 1 and 2 above inside the three-dimensional probability simplex* Δ_{4-1} *as in Figure 2.1. Hint: compare the development in Section 4.2.*

Exercise 2.4. *In the BN represented by the DAG* $\mathcal{G}_1$ *in Figure 2.2 use Proposition 2.4 to check whether the statements* $U \perp\!\!\!\perp Z \mid X$ *and* $U \perp\!\!\!\perp V \mid Z$ *are true.*

Exercise 2.5. *Prove that with locally and globally independent priors of a given BN its Bayes Factor can be expressed as a product of the Bayes Factor of its component model. Calculate this formula explicitly for the example given in Figure 2.2 above where you assume that all variables in the system are binary.*

3

The Chain Event Graph

In this chapter, we introduce the Chain Event Graph as a new graphical statistical model based on an event tree. Section 3.1 slowly builds the formal mathematical foundations for an analysis of the properties of such a model. In particular, we will first outline how the probability tree can be used as a graphical representation of a certain probability space, and how an additional colouring of this tree can capture implicit conditional independence assumptions associated to the depicted events. Coloured probability trees are then called *staged trees* and can be used as graphical representations of discrete parametric statistical models. Just like with the BN, every staged tree thus represents a family of probability distributions for which the depicted assumptions are valid. We often use the staged tree semantics for proving more technical results. For transparency, interpretability and for the design of fast algorithms, we will however tighten the tree graph structure by transforming it into a new object with typically fewer vertices and edges and less redundancy, called the *Chain Event Graph.* This is done in Section 3.2 where a number of examples of transformations from staged trees to Chain Event Graphs and back again illustrate this mechanism. A staged tree and its corresponding CEG are both graphical representations of the same statistical model. In Sections 3.3 and 3.4, we present subclasses of Chain Event Graphs which will be used throughout the book, and compare these with well-known types of Bayesian networks as introduced in the previous chapter. Section 3.5 sets our new theory in the context of similar tree models developed by other authors.

3.1 Models represented by tree graphs

There is a vast literature on graphical models represented by tree structures, starting with models that can be represented using DAGs which are trees (often including hidden variables). For us the most interesting graphs are of tree form but their interpretation is quite different from that of a BN: in particular, in the event-tree framework we develop below, vertices always correspond to events in an underlying probability space—and not to random variables. Every edge then depicts the possibility of transitioning from the donating vertex to the receiving vertex and can be labelled by a corresponding transition probability. This is in contrast to the graphical model from Definition 2.2 where edges represent certain dependencies between the donating and receiving vertex.

3.1.1 Probability trees

We will first introduce some notation from graph theory which enables us to discuss discrete and parametric statistical models represented by *probability trees*. These will include staged tree models (and hence Chain Event Graphs) as a special case.

A finite graph denoted as a pair $\mathcal{T} = (V, E)$ with vertex set V and edge set $E \subseteq V \times V$ is called a *tree* if it is connected and has no cycles. In a *directed* tree, every edge $e = (v, v') \in E$ is a pair of ordered vertices. We call the set of vertices $\mathrm{pa}(v) = \{v' \mid \text{there is } (v', v) \in E\}$ the *parents* of $v \in V$ and $\mathrm{ch}(v) = \{v' \in V \mid \text{there is } (v, v') \in E\}$ the set of *children* of $v \in V$. A vertex without parents is called a *root* of the tree, usually denoted $v_0 \in V$, and vertices without children are called *leaves*. Non-leaves, or inner vertices, are also called *situations*. We use the term *root-to-leaf path* and the symbol λ for a directed and connected sequence of edges $E(\lambda) \subseteq E$ emanating from the root and terminating in a leaf. A *subpath* in a directed tree is then a connected subsequence of a root-to-leaf path. We call a directed tree an *event tree* if all vertices except for one unique root have exactly one parent, each parent has at least two children and the directionality of the graph is such that all edges point away from the root.

When representing a tree model as defined below, an event tree graph depicts all possible unfoldings of events within a population. In particular, every root-to-leaf path then corresponds to an atomic event, depicting one possible history of a unit in the population of interest. Every vertex $v \in V$ denotes a state that a unit following a particular root-to-leaf path might find itself in, and every edge $e = (v, v')$ denotes the possibility of passing from one vertex v to the next v'. Event trees will normally be equipped with labels associated to edges that indicate the probabilities of passing from one vertex to another, as explained in more detail below.

We henceforth denote the set of all root-to-leaf paths of an event tree by $\Lambda(\mathcal{T})$. Two types of subsets of this set will play a central role. For fixed $e \in E$ we define the *edge-centred event* $\Lambda(e) = \{\lambda \in \Lambda(\mathcal{T}) \mid e \in E(\lambda)\}$ as the union of all root-to-leaf paths which pass through that edge. In analogy, we define the *vertex-centred event* $\Lambda(v) \subseteq \Lambda(\mathcal{T})$ as the union of all root-to-leaf paths passing through that vertex. These sets are sometimes called *Moivrean events* [90].

We call a pair $\mathcal{F}(v) = (v, E(v))$ of a vertex $v \in V$ together with its emanating edges $E(v) = \{(v, v') \in E \mid v' \in \mathrm{ch}(v)\}$ a *floret*. Directed root-to-leaf paths passing through two florets induce a natural order $\prec$ on these and on the corresponding vertex-centred events. If there is a root-to-leaf path which passes through v before it passes through v' we will hence say that the vertex v' is *upstream* of v—in symbols $v' \prec v$—and that v is *downstream* of v'.

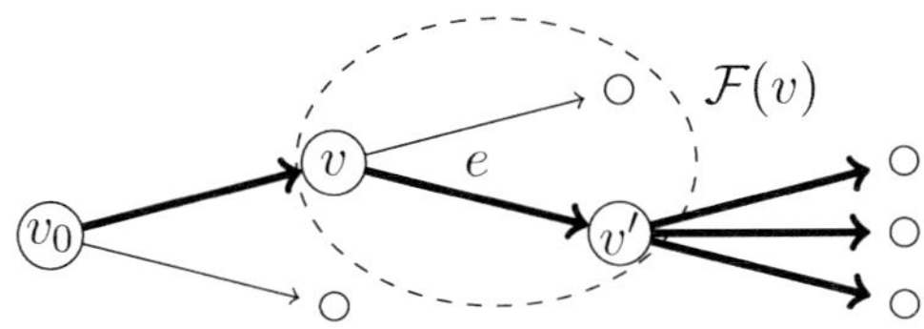

FIGURE 3.1
An event tree $\mathcal{T} = (V, E)$ with root $v_0 \in V$ and situations $v, v' \in V$ which are connected by the edge $e = (v, v') \in E$. The floret $\mathcal{F}(v)$ is encircled. The thick depicted edges correspond to root-to-leaf paths in the edge-centred event $\Lambda(e)$. This is equal to the vertex-centred event $\Lambda(v')$. For simplicity, leaf vertices are often not named.

An illustration of all different concepts introduced above is provided in the event tree depicted in Figure 3.1.

We can now equip event trees with a probability distribution.

Definition 3.1 (Probability tree)**.** *Let $\mathcal{T} = (V, E)$ be an event tree with parameters $\theta(e) = \theta(v, v')$ associated to all edges $e = (v, v') \in E$. We call the vector $\boldsymbol{\theta}_v = \big(\theta(e) \mid e \in E(v)\big)$ of all parameters associated to the same floret a vector of* floret parameters. *If all labels are strictly positive probabilities and the components of all floret parameter vectors sum to unity, so $\theta(e) \in (0, 1)$ and $\sum_{e \in E(v)} \theta(e) = 1$ for all $e \in E$ and all non-leaves $v \in V$, then the pair $(\mathcal{T}, \boldsymbol{\theta}_\mathcal{T})$ of graph $\mathcal{T}$ and vector of all labels $\boldsymbol{\theta}_\mathcal{T} = \big(\theta(e) \mid e \in E\big)$ is called a* probability tree.

We can interpret the label $\theta(v, v')$ on an edge $e = (v, v') \in E$ in a probability tree as the *transition probability* of passing from a situation v to a situation or leaf v' along a root-to-leaf path which contains e. This is also sometimes denoted as the conditional probability $\pi_e(v'|v)$ of 'passing on to v', given arrival at v' in early publications [98, 106]. We will in this and the next chapter mainly think of the labels of an event tree either as parameters specifying the statistical model represented by that tree as in (3.2) or as *primitive* probabilities—similarly to potentials in BNs as in (2.34) [63]. In later chapters we will then present methods to estimate and propagate these parameters and will in this context think of them as having been assigned a fixed meaning in terms of marginal or conditional probabilities, depending on their location in the graph: compare Chapter 5 in particular.

A floret parameter vector $\boldsymbol{\theta}_v = \big(\theta(e) \mid e \in E(v)\big)$ attached to a non-leaf vertex $v \in V$ in a probability tree can be thought of as a vector of all conditional transition probabilities $\theta(e)$ emanating from that situation, $e \in E(v)$. We will show in Section 3.3 below that these are analogous to rows of conditional probability tables of BNs. Section 4.1 then presents how, thanks to this interpretation, random variables associated to vertices can be specified from a tree.

Probability trees are frequently used to visualise how events in a discrete setting might evolve, as we have illustrated throughout Chapter 1 [90]. They are a lot less commonly thought of as graphical statistical models in their own right, in the way that DAGs are widely used as graphical representations of BNs: see again Section 2.3. But this is precisely how we will treat probability trees throughout this book: as representations of certain statistical models.

We briefly repeat the formalism we have developed around probability trees below, referring back to [43, 47] for further more technical details.

We will always denote the product of all primitive probabilities along a root-to-leaf path $\lambda \in \Lambda(\mathcal{T})$ in a probability tree $(\mathcal{T}, \boldsymbol{\theta}_\mathcal{T})$ by

$$p_{\boldsymbol{\theta},\mathcal{T}}(\lambda) = \prod_{e \in E(\lambda)} \theta(e) \tag{3.1}$$

where $\boldsymbol{\theta} = \boldsymbol{\theta}_\mathcal{T}$. It can easily be shown that $p_{\boldsymbol{\theta},\mathcal{T}}$ defines a strictly positive probability mass function on the set of root-to-leaf paths $\Lambda(\mathcal{T})$ of the probability tree. We can then think of every $\lambda \in \Lambda(\mathcal{T})$ representing an atomic event in an underlying probability space, and each of these atomic events will have a positive probability. This assumption is central to our use of models represented by probability trees: we will not depict zero-probability events in an event tree representation of such a model. This enables us to avoid issues related to faithfulness assumptions as present in DAG representations of BN models.

In order to be able to provide a few insightful illustrations of probability tree models in the next section, we will again think of distributions $\boldsymbol{p}_{\boldsymbol{\theta},\mathcal{T}} = \big(p_{\boldsymbol{\theta},\mathcal{T}}(\lambda) \mid \lambda \in \Lambda(\mathcal{T})\big)$ over the atoms represented in a probability tree $(\mathcal{T}, \boldsymbol{\theta}_\mathcal{T})$ as vectors or points in Euclidean space: compare the more general Definition 2.1 of discrete statistical models given in Section 2.2. The set of all vectors which respect the assumptions coded in a given probability tree then defines a parametric statistical model

$$\boldsymbol{M}_{(\mathcal{T},\boldsymbol{\theta}_\mathcal{T})} = \Big\{\boldsymbol{p}_{\boldsymbol{\theta},\mathcal{T}} \mid \boldsymbol{\theta}_v \in \Delta^\circ_{\#E(v)-1} \text{ for all } v \in V\Big\} \tag{3.2}$$

represented by that tree. We call the set in (3.2) a *(probability) tree model* and say that the elements in $\boldsymbol{M}_{(\mathcal{T},\boldsymbol{\theta}_\mathcal{T})}$ *factorise according to* $\mathcal{T}$. This nomenclature is analogous to the one often used in BNs where distributions factorise according to a DAG. We will always index a tree model $\boldsymbol{M}_{(\mathcal{T},\boldsymbol{\theta}_\mathcal{T})}$ by a graphical representation $(\mathcal{T}, \boldsymbol{\theta}_\mathcal{T})$ instead of by the associated parametrisation $\Psi_\mathcal{T} : \boldsymbol{\theta}_\mathcal{T} \mapsto \boldsymbol{M}_{(\mathcal{T},\boldsymbol{\theta}_\mathcal{T})}$. This is because two graphs might be different but still share the same parametrisation. If unambiguous, this is sometimes abbreviated to $\boldsymbol{M}_\mathcal{T} = \boldsymbol{M}_{(\mathcal{T},\boldsymbol{\theta}_\mathcal{T})}$.

Example 3.2 (A coin toss). *Assume we toss a biased coin and then toss a second biased coin if the outcome of the first is* heads. *This toy experiment can be easily depicted by the probability tree* $(\mathcal{T}, \boldsymbol{\theta}_\mathcal{T})$ *given in Figure 3.2. Here, the event tree* $\mathcal{T} = (V, E)$ *has vertex set* $V = \{v_0, v_1, v_2, v_3, v_4\}$ *and edge set*

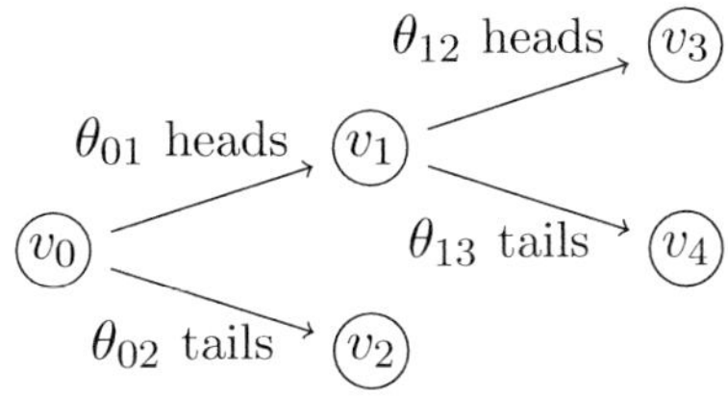

FIGURE 3.2
A probability tree representing the repeated coin toss in Example 3.2.

$E = \{(v_0, v_1), (v_0, v_2), (v_1, v_3), (v_1, v_4)\}$. *The only situations are the root* v_0 *and the inner vertex* v_1*, the leaves are given by* v_2*,* v_3 *and* v_4*. The root-to-leaf paths are* $\lambda_1 = ((v_0, v_1), (v_1, v_3))$*,* $\lambda_2 = ((v_0, v_1), (v_1, v_4))$ *and* $\lambda_3 = ((v_0, v_2))$*. These represent the events* heads, heads *and* heads, tails *and* tails*, respectively.*

The edges in the tree $\mathcal{T}$ *are labelled by the primitive probabilities* $\theta_{01} = \theta(v_0, v_1)$*,* $\theta_{02} = \theta(v_0, v_2)$*,* $\theta_{13} = \theta(v_1, v_3)$ *and* $\theta_{14} = \theta(v_1, v_4)$*. The components of the floret parameter vectors* $\boldsymbol{\theta}_{v_0} = (\theta_{01}, \theta_{02})$ *and* $\boldsymbol{\theta}_{v_1} = (\theta_{13}, \theta_{14})$ *which are attached to the situations are strictly positive and sum to unity.*

The atomic probabilities represented by $(\mathcal{T}, \boldsymbol{\theta}_{\mathcal{T}})$ *are given by the monomials*

$$p_{\boldsymbol{\theta},\mathcal{T}}(\lambda_1) = \theta_{01}\theta_{12},$$
$$p_{\boldsymbol{\theta},\mathcal{T}}(\lambda_2) = \theta_{01}\theta_{14},$$
$$p_{\boldsymbol{\theta},\mathcal{T}}(\lambda_3) = \theta_{02},$$

as in (3.1) where the bold symbol $\boldsymbol{\theta} = \boldsymbol{\theta}_{\mathcal{T}} = (\theta_{01}, \theta_{02}, \theta_{13}, \theta_{14})$ *denotes a vector of all primitive probabilities.*

Now assume that the floret parameter vector attached to the root takes the values $\boldsymbol{\theta}_{v_0} = (0.3, 0.7)$ *and the floret parameter vector of the situation* v_1 *takes the values* $\boldsymbol{\theta}_{v_1} = (0.8, 0.2)$*. These vectors have strictly positive components which sum to one, as required. In particular, the first coin now takes the value* heads *with probability* 0.3 *and* tails *with probability* 0.7*, and the outcomes of the second coin are assigned probabilities* 0.8 *and* 0.2*, respectively. Then the atomic probabilities as above can be calculated as*

$$p_{\boldsymbol{\theta},\mathcal{T}}(\lambda_1) = 0.3 \cdot 0.8 = 0.24,\; p_{\boldsymbol{\theta},\mathcal{T}}(\lambda_2) = 0.3 \cdot 0.2 = 0.06 \text{ and } p_{\boldsymbol{\theta},\mathcal{T}}(\lambda_3) = 0.7.$$

So we can simply read the probability of atomic events off the graph, as products along the respective root-to-leaf paths. If we are interested in calculating the probability of obtaining heads *in the first toss, we calculate this as the probability of the two atomic events* heads, heads *and* heads, tails*. This union is represented in the tree by the vertex-centred event* $\Lambda(v_1) = \{\lambda_1, \lambda_2\}$ *or alternatively by the edge-centred event* $\Lambda((v_0, v_1)) = \Lambda(v_1)$*. Hence,*

$$p_{\boldsymbol{\theta},\mathcal{T}}(\lambda_1) + p_{\boldsymbol{\theta},\mathcal{T}}(\lambda_2) = 0.24 + 0.06 = 0.3 \tag{3.3}$$

which is precisely the probability attached to the first edge labelled heads*. This* *is not a coincidence: in fact, the simple reason for this result is that the floret downstream of* v_1 *has labels which sum to unity, so the probability of* $\Lambda(v_1)$ *is equal to the probability of a unit arriving at* v_1*, via the edge* (v_0, v_1)*. We will see a formal presentation of this observation in Section 4.1.*

It is easy to find many different values of floret parameter vectors, so many different choices of primitive probabilities, which give the same distribution over the depicted atoms as above. In fact, the probability tree model represented by $(\mathcal{T}, \boldsymbol{\theta}_\mathcal{T})$ *is given by all distributions over three atoms which are of the form*

$$\boldsymbol{M}_{(\mathcal{T},\boldsymbol{\theta}_\mathcal{T})} = \big\{(\theta_{01}\theta_{13}, \theta_{01}\theta_{14}, \theta_{02}) \mid \theta_{i1} + \theta_{i2} = 1 \textit{ and } \theta_{i1}, \theta_{i2} \in (0,1), i = 0, 1\big\}$$

as in (3.2).

We will now equip probability trees with an additional graphical property. This can capture certain conditional independence assumptions on distributions which factorise according to these trees. We will then be able to provide expressive illustrations to the concepts introduced above and to develop the Chain Event Graph as a graphical model based on a probability tree.

3.1.2 Staged trees

Probability trees are most interesting when two or more vectors of floret parameters take the same values so that the distributions $p_{\boldsymbol{\theta},\mathcal{T}}$ factorise according to a 'coloured' graph $\mathcal{T}$ which captures these equalities.

Definition 3.3 (Staged tree)**.** *Let* $(\mathcal{T}, \boldsymbol{\theta}_\mathcal{T})$ *with graph* $\mathcal{T} = (V, E)$ *and labels* $\boldsymbol{\theta}_\mathcal{T} = (\boldsymbol{\theta}_v \mid v \in V)$ *be a probability tree. Whenever two floret parameter vectors are equal* $\boldsymbol{\theta}_v = \boldsymbol{\theta}_w$ *up to a permutation of their components we say that their vertices* v *and* w *are in the same* stage*, for* $v, w \in V$*. To every stage we assign one unique colour. If all vertices are either in the same stage or have pairwise different labels then* $(\mathcal{T}, \boldsymbol{\theta}_\mathcal{T})$ *is said to be a* staged tree*.*

A staged tree model *is a probability tree model whose distributions factorise according to a staged tree as in (3.2).*

If in Example 3.2 above we were to toss the same biased coin twice then in the probability tree in Figure 3.2 the vertices v_0 and v_1 would be in the same stage because their attached vectors of floret labels would be equal. In particular, all edges labelled *head* would have the same attached primitive probability and so would all edges labelled *tails*. Compare also Example 4.13 in the next chapter.

In simple words, whenever two vertices are in the same stage and a unit arrives at one of them, the transition probabilities to all children of that vertex will not depend on which of the two vertices the unit is actually in, and will thus not depend on the way that unit took to arrive in that situation. The edge (or transition) probabilities in these stages are thus in a sense independent of their history or location in the tree. We will provide a formal presentation of this type of conditional independence in Section 4.1.

By definition, in staged trees this identification of conditional distributions can always be depicted using colour. In particular, we will always colour all vertices which are in the same stage in the same fashion. This enables us to elegantly communicate all modelling assumptions in a staged tree model in a purely graphical way, making it especially easy for non-experts to read staged trees. In addition, whenever a staged tree arises from an underlying set of problem variables as in Sections 3.3 and 3.4 below, we sometimes use different shapes of vertices in order to stress which vertices correspond to the same random variable: see for instance the running example in Section 5.2. These colourings will also enable us to easily identify stages with the same attached probabilities between different graphs: a property that is particularly useful to the real-world application we develop over Chapter 7.

We will henceforth denote the partition on the vertex set of a staged tree into equivalence classes of vertices which are in the same stage by

$$U_{\mathcal{T}} = \{u \subseteq V \mid v \text{ and } w \text{ are in the same stage for all } v, w \in u\} \tag{3.4}$$

and we call $U_{\mathcal{T}}$ the *stage set* of a staged tree with graph $\mathcal{T}$.

If a distribution factorises according to a coloured tree, its atomic probabilities are given by products of primitive probabilities (3.1) along root-to-leaf paths where labels are identified if two vertices are in the same stage. As a consequence, if we are given a distribution, stage structure—just like conditional independence information in BN models—is *implicitly* provided by the probability mass function. But conversely, we can always specify a staged tree *explicitly* when domain knowledge suggests that certain transition probabilities should be identified. This can then subsequently be translated into a family of probability mass functions which determine a model: see Section 5.1. In particular, a graphical depiction of a given set of atoms together with a stage set are sufficient to identify a staged tree model. Formally, this insight enables us to alternatively think of a staged tree as the pair $(\mathcal{T}, U_{\mathcal{T}})$ rather than $(\mathcal{T}, \boldsymbol{\theta}_{\mathcal{T}})$.

The example below illustrates the advantages of a staged tree representation over a DAG for the same problem [45, 98]. A more detailled comparison of these two types of graphical models is the subject of Section 3.3.

Example 3.4 (An example from biology). *We consider the following simplification of a real system. Suppose a statistical model is designed to explain a possible unfolding of the following events in a cell culture. Initially, a cell finds itself in a benign or hostile environment. The level of activity between cells within this environment might be high or low, and if the environment is hostile then a cell gets damaged and might either survive or die. Surviving cells might make a full or partial recovery. We learn from experts that the level of cell activity is independent of the environment being hostile or benign, that whether or not a cell dies does not depend on its activity and that if a cell does survive then regardless of its history it will fully or partially recover with the same respective probabilities.*

A BN modeller proposes to model this setting using four binary random variables. The state of the environment can then be represented by a random variable X_1 *taking values in a state space* $\mathbb{X}_1 = \{\text{hostile}, \text{benign}\}$*. Cell activity can be measured by* X_2 *with* $\mathbb{X}_2 = \{\text{high}, \text{low}\}$*, viability via* X_3 *with* $\mathbb{X}_3 = \{\text{die}, \text{survive}\}$ *and recovery via* X_4 *with* $\mathbb{X}_4 = \{\text{full}, \text{partial}\}$*. Then the model assumptions above translate into the conditional independence statements* $X_1 \perp\!\!\!\perp X_2$ *and* $X_1, X_2 \perp\!\!\!\perp X_4 \mid X_3$*. These can be represented by the DAG* $\mathcal{G}$ *given below:*

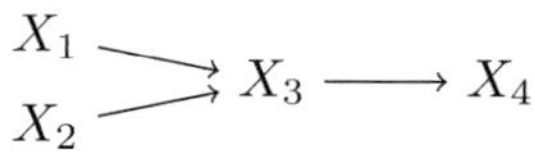

Now, the BN model represented by $\mathcal{G}$ *is not sufficient to accurately capture all assumptions on the variables* $X_1, \ldots, X_4$ *as described by the experts above. In particular, the context-specific constraint that death or survival depends only on the state of the environment cannot be read from the DAG. We further retain redundant information in the product state space* $\mathbb{X} = \mathbb{X}_1 \times \mathbb{X}_2 \times \mathbb{X}_3 \times \mathbb{X}_4$ *of the random variables. For instance, all states* $(\text{benign}, x_2, x_3, x_4) \in \mathbb{X}$ *have probability zero because there is no cell damage in a benign environment and all states* $(x_1, x_2, \text{die}, x_3) \in \mathbb{X}$ *are meaningless because if a unit has died, then there is no recovery possible. It is interesting to note that many real BNs have such sparse probability tables. We can here represent the state space* $\mathbb{X}$ *by a 'degenerate' staged tree denoted* $(\mathcal{T}, \boldsymbol{\theta}_{\mathcal{T}})_{\mathcal{G}}$ *in Figure 3.3a which includes edges with zero probability. Details on how we obtained this tree representation can be found in Section 3.3.*

The graphical symmetries apparent in the figure above are typical for staged trees that enjoy an alternative DAG representation: all paths are of the same length and the stage structure depends on the distance of a vertex from the root. Proposition 3.10 states this observation as a more general result. From the above, we notice also that the bottom right part of the graph in Figure 3.3a does not contain any valuable information. $(\mathcal{T}, \boldsymbol{\theta}_{\mathcal{T}})_{\mathcal{G}}$ *thus nicely visualises how much redundant information there is in the BN model. We could improve this representation using a context-specific BN rather than a BN—in fact, such a solution is often used in practice. However, there is a strong case here for using a staged tree model instead. Of course when data is available then the model selection methods presented in Section 6.2 will help us to formally compare two competing models.*

We thus propose to model the biologists' description of the problem above using the staged tree $(\mathcal{T}, \boldsymbol{\theta}_{\mathcal{T}})$ *in Figure 3.3b. This new representation is far more expressive than the DAG* $\mathcal{G}$ *and less cluttered than the corresponding tree* $(\mathcal{T}, \boldsymbol{\theta}_{\mathcal{T}})_{\mathcal{G}}$*, whilst conveying* all *model information: all edges with probability zero and all unfoldings which are meaningless have been deleted, and the stage structure visually expresses the conditional independence assumptions given above. In particular, we identify the floret parameter vectors* $\boldsymbol{\theta}_{v_1} = \boldsymbol{\theta}_{v_2}$ *because the activity between cells does not depend on the environment (so the transition probabilities from* v_1 *and* v_2 *are the same),* $\boldsymbol{\theta}_{v_3} = \boldsymbol{\theta}_{v_4}$ *because death or survival*

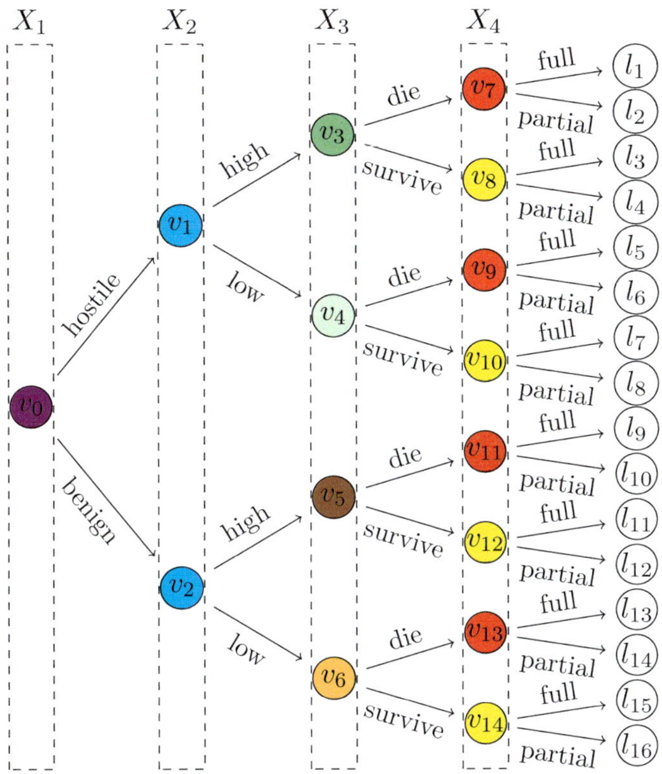

(a) A degenerate staged tree $(\mathcal{T}, \boldsymbol{\theta}_{\mathcal{T}})_{\mathcal{G}}$ depicting the state space of a BN with DAG $\mathcal{G}$.

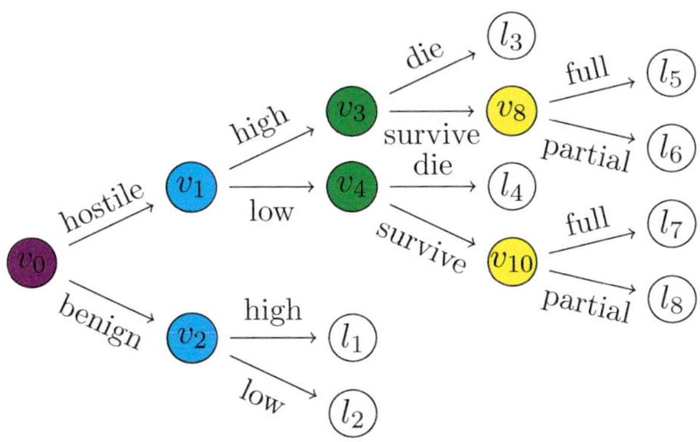

(b) A staged tree $(\mathcal{T}, \boldsymbol{\theta}_{\mathcal{T}})$ representing context-specific information.

FIGURE 3.3
Two staged trees representing alternative models developed for the problem in Example 3.4.

does not depend on cell activity given the environment is hostile, and $\boldsymbol{\theta}_{v_8} = \boldsymbol{\theta}_{v_{10}}$ *because the chances of recovery of a surviving cell are independent of the history of that cell. So the stage set of this tree equals* $U_{\mathcal{T}} = \{u_0 = \{v_0\}, u_1 = \{v_1, v_2\}, u_2 = \{v_3, v_4\}, u_3 = \{v_8, v_{10}\}\}$*. We colour to all vertices in* u_0 *purple, all vertices in* u_1 *blue, all vertices in* u_2 *green, and all vertices in* u_3 *yellow. The leaves* $l_1, \ldots, l_8$ *are all in the same stage but have no attached probability distributions so they are not coloured. The staged tree* $(\mathcal{T}, \boldsymbol{\theta}_{\mathcal{T}})$ *then represents a statistical model which is not only of lower dimension—*$8-1=7$ *rather than* $16-1=15$*—and has a simpler graphical representation than the BN but is also a much more accurate representation of the situation at hand. In this type of* asymmetric *modelling context where a large amount of context-specific information is present, the use of staged tree models can thus be highly advantageous.*

Despite the apparent advantages of the staged tree in Figure 3.3b of Example 3.4 over a BN modelling the same problem, we can also guess from the Figures 3.3a and 3.3b that, even in moderately sized problems, graphical model representations based on event trees quickly become huge. This is one of the main motivations for us to introduce a more compact graphical description based on an underlying staged tree. This new object has then both representational and computational advantages whilst retaining the expressiveness of a staged tree.

3.2 The semantics of the Chain Event Graph

We will now define a new graph based on a staged tree where many of the edges with duplicated colours and their associated parent vertices are either deleted or merged. There is a one-to-one correspondence between the coloured staged tree and this new graph so that despite the new representation having many less vertices and edges, we can still infer the same information from both.

Thus let $(\mathcal{T}, \boldsymbol{\theta}_{\mathcal{T}})$ be a staged tree. We first develop notation for capturing the observation that often stage structure can in a sense be repetitive along a collection of root-to-leaf paths. We will hence in the following denote by $\mathcal{T}(v) \subseteq \mathcal{T}$ the event tree which is rooted at $v \in V$ and whose root-to-leaf paths are precisely the v-to-leaf paths in $\mathcal{T} = (V, E)$. If $\mathcal{T}(v)$ has edge labels inherited from $\mathcal{T}$, we say that the pair $(\mathcal{T}(v)\, \boldsymbol{\theta}_{\mathcal{T}(v)})$ is a *(probability) subtree* of $(\mathcal{T}, \boldsymbol{\theta}_{\mathcal{T}})$. We will then say that two vertices $v, v' \in u$ which are in the same stage $u \in U_{\mathcal{T}}$ are also in the same *position* if their subtrees $(\mathcal{T}(v), \boldsymbol{\theta}_{\mathcal{T}(v)})$ and $(\mathcal{T}(v'), \boldsymbol{\theta}_{\mathcal{T}(v')})$ have the same graph and have the same sets of edge labels, after an identification of their respective vertices and edges.

This implies that for two vertices in the same position all downstream stage structure is the same, and not only the immediate staging. For instance, in Figure 3.3a the vertices v_1 and v_2 are in the same stage as well as in the same position. In Figure 3.3b, the vertices v_3 and v_4 are also in the same position but for example v_1 and v_2 are in the same stage without being in the same position. See also Example 3.6 below.

Now, the position relation induces a new partition $W_{\mathcal{T}}$ on the vertex set of a staged tree. This is by construction coarser than the one induced by the stage relation in (3.4). Hereby, all leaves are trivially in the same position, denoted $w_\infty \in W_{\mathcal{T}}$ and called a *sink* node, and the root is the only element in a position $w_0 \in W_{\mathcal{T}}$, also called the *root*.

We can thus construct the following graph:

Definition 3.5 (Chain Event Graph). *Let $(\mathcal{T}, \boldsymbol{\theta}_{\mathcal{T}})$ be a staged tree with graph $\mathcal{T} = (V, E)$. Denote the set of positions of this tree by $W_{\mathcal{T}}$. We build a new labelled graph $(\mathcal{C}(\mathcal{T}), \boldsymbol{\theta}_{\mathcal{T}})$ as follows:*

$\mathcal{C}(\mathcal{T}) = (W, F)$ is a graph with vertex set $W = W_{\mathcal{T}}$ given by the set of positions in the underlying staged tree. Every position inherits its colour from the staged tree. F is a set of possibly multiple edges between these vertices with the following properties. If there exist edges $e = (v, v'), e' = (w, w') \in E$ and v, w are in the same position then there exist corresponding edges $f, f' \in F$. If also v', w' are in the same position, then $f = f'$. The labels $\theta(f)$ of edges $f \in F$ in the new graph are inherited from the corresponding edges $e \in E$ in the staged tree $(\mathcal{T}, \boldsymbol{\theta}_{\mathcal{T}})$.

We call the labelled graph $(\mathcal{C}(\mathcal{T}), \boldsymbol{\theta}_{\mathcal{T}})$ the Chain Event Graph (CEG) *with underlying staged tree $(\mathcal{T}, \boldsymbol{\theta}_{\mathcal{T}})$.*

When drawing a CEG, we often emphasise the special status of the sink node by not assigning a colour to it and not encircling it as other vertices in the graph. This is because whilst all units in a system represented by a CEG eventually arrive in a final state, that final state is not the same for every unit: for instance, not all units will be dead in Example 3.4. Compare Figure 3.4 below.

In the following, we will often also call the graph $\mathcal{C}$ a CEG and, if unambiguous, we will shorten notation to $\mathcal{C} = (\mathcal{C}(\mathcal{T}), \boldsymbol{\theta}_{\mathcal{T}})$ without explicitly referring to labels or the underlying staged tree.

In particular, because the CEG $(\mathcal{C}(\mathcal{T}), \boldsymbol{\theta}_{\mathcal{T}})$ is unique to the staged tree $(\mathcal{T}, \boldsymbol{\theta}_{\mathcal{T}})$, we can again identify a CEG from the pair $(\mathcal{T}, U_{\mathcal{T}})$ of event tree and stages. Throughout this book, we will use the calligraphic letter $\mathcal{C}$ exclusively when referring to a graphical representation. We will use the bold letter $\boldsymbol{C}$ for a class of models $\boldsymbol{M}_{\mathcal{C}} = \boldsymbol{M}_{(\mathcal{C}(\mathcal{T}), \boldsymbol{\theta}_{\mathcal{T}})}$ represented by CEGs, so $\boldsymbol{C} = \{\boldsymbol{M}_{\mathcal{C}} \mid \mathcal{C} \text{ is a CEG}\}$. Such a *CEG model* $\boldsymbol{M}_{\mathcal{C}}$ is simply a model whose distributions factorise according to the staged tree underlying the CEG $\mathcal{C}$: this definition is justified by Theorem 1 below. In the next two sections, we will investigate some important subclasses of the class of all CEG models.

Throughout the book, we will extend all vocabulary which we have introduced for event trees at the beginning of this section—root-to-leaf/sink paths,

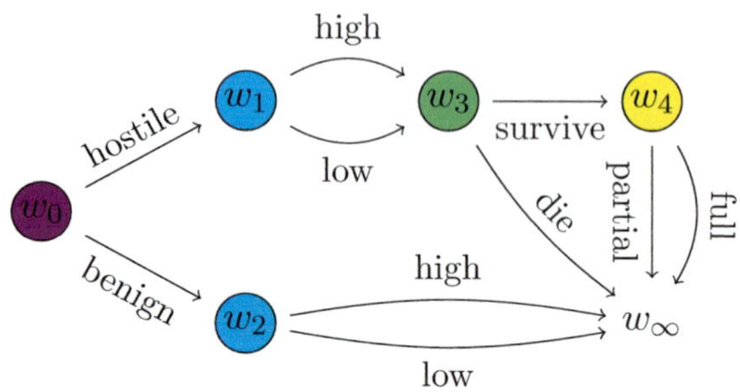

FIGURE 3.4
A CEG $(\mathcal{C}(\mathcal{T}), \boldsymbol{\theta}_{\mathcal{T}})$ whose underlying staged tree $(\mathcal{T}, \boldsymbol{\theta}_{\mathcal{T}})$ is depicted in Figure 3.3b. See Examples 3.4 and 3.6.

edge- or vertex-centred events, florets and others—to CEGs. In particular, just like staged trees also CEGs have stages and can have two identically coloured positions in the same stage. We illustrate this below.

Example 3.6 (Constructing a CEG)**.** *Consider again the staged tree* $(\mathcal{T}, \boldsymbol{\theta}_{\mathcal{T}})$ *from Example 3.4. Here, the stage set* $U_{\mathcal{T}}$ *contains precisely the following elements: the root* $u_0 = \{v_0\}$ *coloured purple, the stage* $u_1 = \{v_1, v_2\}$ *coloured blue,* $u_2 = \{v_3, v_4\}$ *coloured green,* $u_3 = \{v_8, v_{10}\}$ *coloured yellow and the leaves* $u_\infty = \{l_1, l_2, l_3, \ldots, l_8\}$ *in Figure 3.3b. Note that the root is the only trivial stage and that all leaves are in the same stage.*

The positions in $W_{\mathcal{T}}$ *are then the root* $w_0 = u_0$*,* $w_1 = \{v_1\}$*,* $w_2 = \{v_2\}$*,* $w_3 = u_2$*,* $w_4 = u_3$ *and the sink* $w_\infty = u_\infty$*. Here,* v_1 *and* v_2 *are the only vertices which are in the same stage but not in the same position. As a consequence,* w_1 *and* w_2 *are in the same stage in the CEG.*

Thus, we can now construct the CEG $(\mathcal{C}(\mathcal{T}), \boldsymbol{\theta}_{\mathcal{T}})$ *corresponding to* $(\mathcal{T}, \boldsymbol{\theta}_{\mathcal{T}})$ *with vertex set* $W = W_{\mathcal{T}}$ *and edges with labels inherited from* $(\mathcal{T}, \boldsymbol{\theta}_{\mathcal{T}})$*. This graph—which is a lot less repetitive than the underlying staged tree—is given in Figure 3.4. Importantly, this CEG inherits the colouring of its staged tree such that vertices with the same meaning and the same transition probabilities can be easily identified across both representations of the same model.*

Despite the graphical simplicity of the CEG we can still read the same unfolding of events from the set of root-to-sink paths of $(\mathcal{C}(\mathcal{T}), \boldsymbol{\theta}_{\mathcal{T}})$ *as we could from* $(\mathcal{T}, \boldsymbol{\theta}_{\mathcal{T}})$*.*

The CEG is a much more compact representation of a staged tree because every structure of the underlying coloured tree that is repetitive, and in this sense redundant, is merged into a single vertex or edge. Colours in this new graph are inherited from positions with the same meaning in the staged tree. The primitive and atomic probabilities are hereby by construction preserved, as are the identifications of root-to-leaf/sink paths and atoms. We thus find:

Theorem 1 (Staged trees and CEGs)**.** *A staged tree* $(\mathcal{T}, \boldsymbol{\theta}_{\mathcal{T}})$ *and its CEG*

$(\mathcal{C}(\mathcal{T}), \boldsymbol{\theta}_\mathcal{T})$ *are graphical representations of the same parametric statistical model* $\boldsymbol{M}_{(\mathcal{T},\boldsymbol{\theta}_\mathcal{T})} = \boldsymbol{M}_{(\mathcal{C}(\mathcal{T}),\boldsymbol{\theta}_\mathcal{T})}$.

A proof is provided in [98]. A consequence of this proposition is that the staged tree and its corresponding CEG can be used interchangeably to represent the same model. An upshot is that all results we find in this book that apply to one class of coloured graphs also automatically apply to the other.

In particular, just like a staged tree, a CEG is a purely graphical representation of a model which is able to embody in its coloured graph information about the underlying probability distribution. This information includes the number of atoms of the sample space, the levels of all associated random variables if the underlying staged tree arises from a set of problem variables, logical constraints on state spaces, equalities of marginal or conditional distributions and a certain order of events. It is easier to handle than a staged tree representation and yields a more compact problem description. Thanks to this graphical compactness, models represented by CEGs are open to a number of fast and efficient statistical inference techniques: we will discuss a vast range of methodologies in Chapters 5 and 6 as well as Chapter 8. Many of these are then illustrated in Chapter 7.

Before we go on to provide first examples of CEGs, let us briefly consider an explanation for the term 'Chain Event Graph'.

Chain Event Graphs were named in 2005. Obviously they defined a class of graphical models analogous to Bayesian networks or chain graphs. But the fact that they were built around an event tree—rather than the latter two classes mentioned which were built with vertices representing variables—suggested that *event* should appear somewhere in the name. The term chain is less obvious. In our first semantic we identified two situations in the same stage by an undirected edge. This then gave us a graph which was a mixed graph—a property shared by the chain graph [63]. In addition, propagation algorithms and interrogation methods we had developed could be seen as 'rubbing out' all these undirected edges, analogous to the moralisation steps in the early junction tree algorithms [56, 102] except there undirected edges were added instead of being omitted. Since there was a close link between these steps and embedding the models in a chain representation the term *chain* seemed a good one.

Subsequently, it was discovered that the use of colour was a much better way of depicting stage structure. This was for two reasons. First, if there were more than a few positions in a given stage then the corresponding CEG became very cluttered and difficult to read. Instead assigning the same colour to two situations in the same stage kept the depiction much cleaner. Secondly, we realised that if the CEG had no obvious collection of random variables associated with it then to unambiguously represent the stage structure we also had to label the edges of the graph to identify which probabilities were being identified with each other. But we could do this formally by colouring its edges. So belatedly we realised that what we were dealing with was a coloured

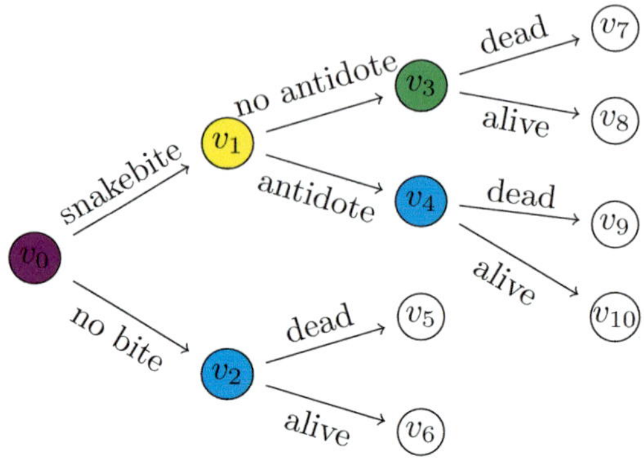

(a) A staged tree representing the snakebite model.

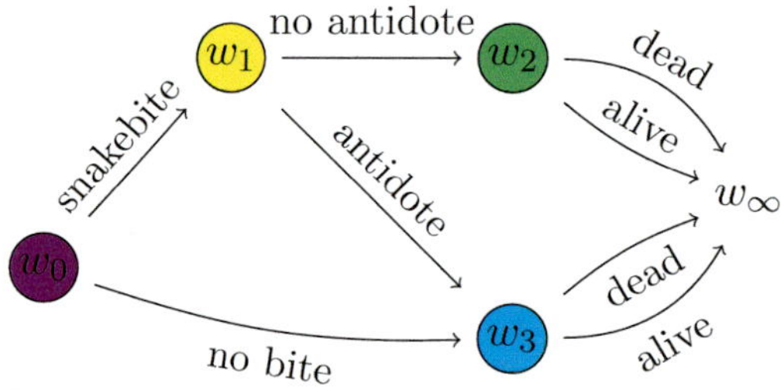

(b) A CEG representing the snakebite model.

FIGURE 3.5
Two graphical representations for Example 3.7.

directed graph not a mixed graph—but the term 'chain' had by then been established and so remains to this day.

Stages in a probability tree often have a very intuitive interpretation in a modelling context, in the sense that vertex-centred events whose vertices are in the same stage can often be given a very similar meaning. This meaning then translates seamlessly into CEG semantics. We have seen a first occurrence of this in Example 3.2 where the introduction of a stage would correspond to tossing the same coin twice. Another toy example below illustrates how to interpret stage structure in a slightly more complex context.

Example 3.7 (A snakebite)**.** *Consider the following setting. An explorer in a jungle is confronted with a snake. A snakebite is not necessarily lethal but might cause serious health issues. The explorer might carry an antidote which if applied will neutralise the snake venom. We want to model this problem in order to assess whether the explorer is alive in a certain situation.*

The staged tree in Figure 3.5a represents the setting above, depicting all six different atomic events. At each situation, the progress through the tree is split into unfoldings given by bite *or* no bite, antidote applied *or* not, dead *or* alive. *Because an application of the antidote would not make sense if not bitten, this staged tree does not necessarily have equivalent unfoldings along different root-to-leaf paths. The known effectiveness of the antidote ensures that the respective probabilities of life and death are the same in the situation 'bitten and antidote applied' and the situation 'not bitten'. The situations* v_2 *and* v_4 *are thus in the same stage* $u_1 = \{v_2, v_4\}$. *We can interpret this stage as 'safe' and the alternative stage* $u_2 = \{v_3\}$ *as 'endangered'. Figure 3.5b shows the CEG representation of this staged tree with positions* $w_2 = u_2$ *and* $w_3 = u_1$ *which have the same respective interpretation.*

This staged tree and CEG represent another typical example where a model could not have been without loss represented by a DAG. In particular, even though this model is incredibly simple its stage structure does not follow any symmetries within the graph.

We will enhance our analysis of the above example in Exercise 3.1 below, investigating different ways to represent the same problem and drawing out advantages of each of these representations.

Over the remainder of this chapter we will investigate the class of CEG models in relation to other established classes of graphical models. In this and the later development, we will often use the staged-tree semantics for illustrating more technical results and we will use the CEG in computations and inference.

3.3 Comparison of stratified CEGs with BNs

Following the development in Section 3.1, we can specify a CEG, or in fact any parametric model as introduced in Definition 2.1, without relying on an underlying set of random variables. This implies a great flexibility of CEG models which will be exploited throughout this book. However, sometimes a problem is naturally defined through the relationships between a set of prespecified random variables. This is particularly the case when we are performing model selection over a wide class of CEGs in Section 6.3 where inferential techniques concern a plausibly prespecified set of random variables. When this is so, we often end up analysing so-called 'stratified' CEGs whose semantics enable us to make use of the extra information coded in the underlying product state space. We will hence show in this section how to construct such a stratified CEG representation for a given BN model.

Let always $\boldsymbol{X} = (X_1, \ldots, X_m)$ denote a vector of discrete random variables which takes values in a product state space, $\mathbb{X} = \mathbb{X}_1 \times \ldots \times \mathbb{X}_m$, for some $m \in \mathbb{N}$, and which is measurable with respect to a given positive probability measure P. We always assume that $P = P_{\boldsymbol{\theta}}$ is parametrised using a vector

$\boldsymbol{\theta} \in \Theta$ taking real values $\Theta \subseteq \mathbb{R}^d$, $d \in \mathbb{N}$, and that atomic probabilities are of monomial form:

$$P_{\boldsymbol{\theta}}(\boldsymbol{X} = \boldsymbol{x}) = \prod_{i=1}^{k} \theta(\boldsymbol{x}_{A_i}) \quad \text{for all } \boldsymbol{x} \in \mathbb{X} \tag{3.5}$$

where $\boldsymbol{x}_{A_i}$ denotes the subvector $(x_j \mid j \in A_i)$ in the product space $\mathbb{X}_{A_i} = \times_{j \in A_i} \mathbb{X}_j$, for index sets $A_i \subseteq \{1, \ldots, m\}$, $i = 1, \ldots, k$ and $k \in \mathbb{N}$. The parameters above can be thought of as *potentials* or *kernels* of a probability mass function which can be interpreted as conditional or marginal probabilities just like in Section 2.3 [63]. Their respective meaning will depend on their location within the graph we build below.

If the probability mass function (3.5) is strictly positive then the state space $\mathbb{X}$ can easily be embedded into the set of root-to-leaf paths $\Lambda(\mathcal{T})$ of a probability tree. The labels of this tree are then precisely given by the potentials above. We show below how this tree can be constructed. Note however that if there are zeros in the conditional probability tables of the variables above then these will translate into edges with zero probability, and hence atoms with probability zero, in the tree representation. In this case we thus build a 'degenerate' probability tree as in Example 3.4.

Given (3.5), we can draw an event tree $\mathcal{T}$ with a root-vertex v_0 that corresponds to the random variable $\boldsymbol{X}_{A_1}$ and which has one emanating edge $e(\boldsymbol{x}_{A_1})$ for every state $\boldsymbol{x}_{A_1} \in \mathbb{X}_{A_1}$ of that variable. Downstream of these edges we then add a vertex $v(\boldsymbol{x}_{A^i})$ to the tree for every random variable $\boldsymbol{X}_{A_i} \mid \boldsymbol{X}_{A^i} = \boldsymbol{x}_{A^i}$ conditional on its ancestors $A^i = \bigcup_{j=1}^{i-1} A_j$, $i = 2, \ldots, k$. To these we attach emanating edges for every element in the state space of that conditional variable, just like for the root. Every floret is thus constructed to have a number of emanating edges equal to the cardinality of the state space of its associated random variable, $\#E(v(\boldsymbol{x}_{A^i})) = \#\mathbb{X}_{A_i}$. So for all vertices which lie at the same distance from the root—or for all vertices on the same *level* of the tree—their associated random variables $\boldsymbol{X}_{A_i}$ are the same up to conditioning on the upstream variables $\boldsymbol{X}_{A^i} = \boldsymbol{x}_{A^i}$, for some $\boldsymbol{x}_{A^i} \in \mathbb{X}_{A^i}$, $i = 2, \ldots, k$. Now the graph $\mathcal{T}$ represents the state space $\mathbb{X}$.

We then attach labels to this event tree given by the parametrisation in (3.5) in such a way that the edge $e(\boldsymbol{x}_{A_i})$ has an attached label $\theta(\boldsymbol{x}_{A_i})$. As a consequence, the floret emanating from the vertex $v(\boldsymbol{x}_{A_i})$ has attached the conditional probabilities $P_{\boldsymbol{\theta}}(\boldsymbol{X}_{A_i} = \boldsymbol{x}_{A_i} \mid \boldsymbol{X}_{A^i} = \boldsymbol{x}_{A^i})$ of all $\boldsymbol{x}_{A^i} \in \mathbb{X}_{A^i}$. Every floret thus depicts the state space of a conditional random variable and its floret parameter vector is the vector of conditional probabilities, or a row in the conditional probability table of that random variable. Then by construction, every root-to-leaf path λ in the probability tree $(\mathcal{T}, \boldsymbol{\theta}_{\mathcal{T}})$ is identified with one atom $\boldsymbol{x} \in \mathbb{X}$ in the underlying product state space and its associated product of primitive probabilities equals the probability of that atom: so $p_{\boldsymbol{\theta},\mathcal{T}}(\lambda) = P_{\boldsymbol{\theta}}(\boldsymbol{X} = \boldsymbol{x})$ if λ represents $\boldsymbol{x}$.

Definition 3.8 ($\boldsymbol{X}$-compatible). *We call a (possibly degenerate) probability tree constructed as above an* $\boldsymbol{X}$-compatible *tree. The CEG constructed from an* $\boldsymbol{X}$*-compatible probability tree is also called* $\boldsymbol{X}$-compatible.

By construction, the representation constructed above is highly dependent on the chosen parametrisation. For instance, if in (3.5) the random variables come from a BN model then we might have that $A_i = \{1, \ldots, i-1\}$ for all $i = 1, \ldots, k$. In this case, we will build a tree that has one single component random variable X_i (conditional on its ancestors) associated to every vertex and the parametrisation is as explicitly given in (2.24). But if instead the parametrisation in (3.5) is based on the cliques of a decomposable BN then we would have a whole clique associated to every vertex, conditional on its separators: precisely as given in (2.34). We illustrate this in Example 3.11 below.

In the construction of an $\boldsymbol{X}$-compatible tree above we have assumed that we are given a probability measure and, implicit in this representation, an ordering of the random variables in the vector $\boldsymbol{X}$. Of course, if there is no such given ordering, we can construct a whole class of $\boldsymbol{X}$-compatible staged trees, all depicting the same state space. As is the case with DAGs, often there are a number of valid orderings for $\boldsymbol{X}$-compatible staged trees which represent the same model whilst other orderings would violate certain model constraints imposed by the directionality of the graphical representation. We will see examples of this below and will provide a more in-depth analysis in Section 4.2. In particular, we then provide conditions under which staged trees (and $\boldsymbol{X}$-compatible staged trees) represent the same model.

Throughout Chapter 6 we will need to distinguish between $\boldsymbol{X}$-compatible trees which follow one given ordering and those which do not. To make this distinction, in the case of a fixed order we will use the term $\boldsymbol{X}(I)$-*compatible* where the notation $\boldsymbol{X}(I)$ indicates a random vector whose components $X_1, \ldots, X_m$ are permuted into an order $I = (i_1, i_2, \ldots, i_m)$ where $\{i_1, \ldots, i_m\} = \{1, \ldots, m\}$. In the corresponding tree, the random variables will then also be ordered as $X_{i_1} \prec X_{i_2} \prec \ldots \prec X_{i_m}$. Whenever we abstain from the rather strong constraint that a set $\boldsymbol{\mathcal{X}} = \{X_1, \ldots, X_m\}$ of given random variables is a priori ordered, we will talk about $\boldsymbol{\mathcal{X}}$*-compatible* trees and assume that such a tree is $\boldsymbol{X}$-compatible with any ordering of the components of $\boldsymbol{\mathcal{X}}$. The set of all $\boldsymbol{\mathcal{X}}$-compatible staged trees will be denoted by the symbol $\mathcal{T}_{\boldsymbol{\mathcal{X}}}$. In particular, $\mathcal{T}_{\boldsymbol{\mathcal{X}}}$ contains all $\boldsymbol{X}$-compatible trees for any ordered vector $\boldsymbol{X}$ of random variables in the set $\boldsymbol{\mathcal{X}}$. We will repeat the necessary notation in Section 6.3.

In practice, a preassigned set of random variables as above is often equipped with some extra conditional independence assumptions—as is the case in BNs. These can be implicitly coded in the distribution (3.5) but of course they can also be very explicitly represented in the colouring of an $\boldsymbol{X}$-compatible probability tree. In particular, every conditional independence assumption and every context-specific independence assumption of the type

$$P_{\boldsymbol{\theta}}(\boldsymbol{X}_A \mid \boldsymbol{X}_{A'} = \boldsymbol{x}_{A'}) = P_{\boldsymbol{\theta}}(\boldsymbol{X}_A \mid \boldsymbol{X}_{A'} = \boldsymbol{y}_{A'}) \tag{3.6}$$

for some $\boldsymbol{x}_{A'} \neq \boldsymbol{y}_{A'}$ and index sets $A' \subseteq A$ as in the notation above, translates into an identification of primitive probabilities $\theta(\boldsymbol{x}_A) = \theta(\boldsymbol{y}_A)$ on an $\boldsymbol{X}$-compatible representation of the model. Because we associate each vertex to a conditional random variable, varying over $\boldsymbol{X}_A = \boldsymbol{x}_A$ in (3.6) does then not only identify single edge labels but vectors of primitive probabilities associated to the same floret. So $\boldsymbol{X}$-compatible staged trees can code the exact same assumptions as the DAG for a BN model *and* can code all assumptions implicit in a context-specific BN model in a purely graphical fashion.

Whenever the distributions underlying an $\boldsymbol{X}$-compatible staged tree come from such a model, the colouring of this tree is very symmetric. We will state this explicitly below and will here introduce a concept that will become important in all model selection algorithms in Section 6.3.

Definition 3.9 (Stratified CEGs). *A staged tree is called* stratified *if all vertices which are in the same stage are also at the same distance of edges from the root. The corresponding CEG is then also called stratified or for short an* SCEG.

In stratified $\boldsymbol{X}$-compatible staged trees and CEGs, by construction the colouring simply identifies rows of conditional probability tables of the underlying random variables. In particular, the class of all (context-specific) BNs is thus a subclass of the class of CEG models. Models with these types of constraints are now widely used in BN modelling, especially when the domain of application is large: see Chapter 2.

It is easy to check that by construction, the following result is true.

Proposition 3.10 ($\boldsymbol{X}$-compatible staged trees). *Let $\boldsymbol{X} = (X_1, \ldots, X_m)$ be a vector of random variables taking values in a product state space with $n \geq 2^m$ elements. Assume $\boldsymbol{X}$ to be measurable with respect to a measure $P_{\boldsymbol{\theta}}$ as in (3.5). Let $(\mathcal{T}, \boldsymbol{\theta}_{\mathcal{T}})$ be an $\boldsymbol{X}$-compatible staged tree with $\boldsymbol{\theta}_{\mathcal{T}} = \boldsymbol{\theta}$. The following are true:*

1. *The event tree $\mathcal{T}$ has n root-to-leaf paths.*
2. *Every root-to-leaf path of $\mathcal{T}$ is of the same length $k \leq m$.*
3. *All vertices in $\mathcal{T}$ which are at the same distance from the root also have the same number of emanating edges.*
4. *For any (context-specific) conditional independence assumptions on $\boldsymbol{X}$, the staged tree $(\mathcal{T}, \boldsymbol{\theta}_{\mathcal{T}})$ is stratified.*

The proposition above summarises what we have already noted in the earlier Example 3.4: staged tree representations of an underlying product state space reflect the symmetries in that space. In addition, BN-type conditional independence assumptions and context-specific conditional independences on the problem variables translate into constraints which are imposed only on

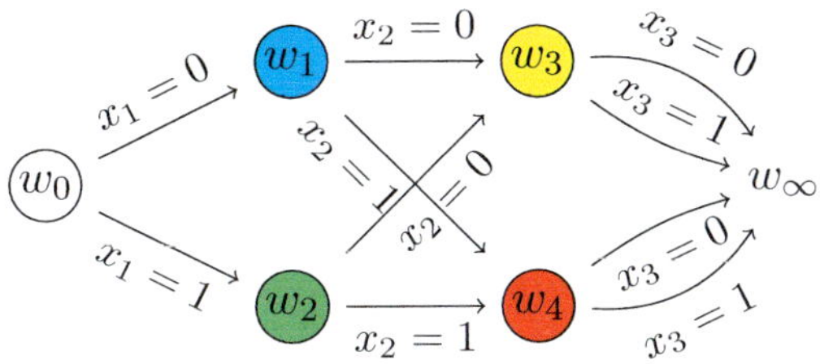

(a) An (X_1, X_2, X_3)-compatible CEG.

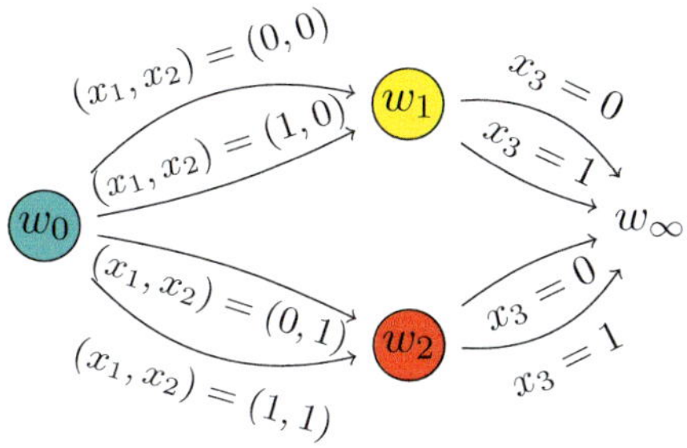

(b) An alternative $(X_{\{1,2\}}, X_3)$-compatible CEG with a different parametrisation.

FIGURE 3.6
Two different $\boldsymbol{X}$-compatible CEGs representing the same model in Example 3.11. The colouring of positions can be translated across these two representations.

certain levels of the underlying tree. These properties naturally extend to an $\boldsymbol{X}$-compatible CEG representation.

In all examples below we will assume random variables to be binary. This constraint is imposed merely to keep our illustrations small—all results naturally apply to random variables with an arbitrary number of states. In Exercise 3.3, the reader will be asked to fill in the details of some of the developments presented below.

Example 3.11 (Conditional independence). *Let $\boldsymbol{X} = (X_1, X_2, X_3)$ be a vector of binary random variables and assume that X_1 is independent of X_3 given X_2, in symbols $X_1 \perp\!\!\!\perp X_3 \mid X_2$. As a BN model, we can represent this conditional independence assumption by either of the three DAGs*

$$\mathcal{G}_1 : X_1 \to X_2 \to X_3, \quad \mathcal{G}_2 : X_1 \leftarrow X_2 \to X_3 \quad \textit{or} \quad \mathcal{G}_3 : X_1 \leftarrow X_2 \leftarrow X_3$$

or the undirected graph $X_1 — X_2 — X_3$.

We will analyse staged tree representations for possible factorisations according to two of these representations. Let thus

$$p(\boldsymbol{x}) = p_1(x_1)p_2(x_2|x_1)p_3(x_3|x_2) \quad \textit{and} \quad p_{\boldsymbol{\theta}}(\boldsymbol{x}) = \theta(x_1, x_2)\theta(x_2, x_3)$$

be the factorisations according to $\mathcal{G}_1$ and according to the undirected graph, here given as a product of potentials, respectively. Always, $\boldsymbol{x} = (x_1, x_2, x_3)$ *denotes one element in the product state space* $\mathbb{X} = \{0, 1\}^3$. *First we observe that the conditional independence assumptions of the BN model are implicit in both of the factorisations above: the probability of* $X_3 = x_3$ *does not depend on the value* $X_1 = x_1$ *if we are given that* $X_2 = x_2$. *Explicitly, fixing* x_2 *and* x_3, *we can write*

$$p_3(x_3|x_1, x_2) = p_3(x_3|x_1', x_2) \quad \text{and} \quad \theta(x_1, x_2, x_3) = \theta(x_1', x_2, x_3)$$

for all $x_1, x_1' = 0, 1$.

Following the procedure outlined at the beginning of this section, we can now draw staged trees with eight root-to-leaf paths, one for every element in $\{0, 1\}^3$, *and colour these using the explicit constraints above. This is left as an exercise to the reader. The resulting* $\boldsymbol{X}$*-compatible CEGs are depicted in Figures 3.6a and 3.6b, respectively.*

The CEG in Figure 3.6a depicts edges associated to the states of X_1 *first, followed by those corresponding to* X_2, *followed by* X_3. *The CEG in Figure 3.6b acknowledges that there are no constraints between* X_1 *and* X_2 *and depicts edges associated to the joint variable* $X_{\{1,2\}} = (X_1, X_2)$ *first, followed by* X_3. *We can see here that different factorisation of the same probability mass function, even though they code the same conditional independence assumptions, can give rise to different graphical representations of the same model. Just like* $\mathcal{G}_1$, $\mathcal{G}_2$ *and* $\mathcal{G}_3$ *are representations of the same BN model, the CEGs in Figures 3.6a and 3.6b are representations of this same statistical model: see Section 4.2 for details on this observation.*

Example 3.12 (Independence). *Let* $m \in \mathbb{N}$ *and let* $\boldsymbol{X} = (X_1, \ldots, X_m)$ *be a vector of binary and mutually independent random variables,* $\perp\!\!\!\perp_{i=1}^{m} X_i$.

As a BN model, this independence assumption would be represented by a DAG with m *vertices and no edges. The probability mass function*

$$p_{\boldsymbol{\theta}}(\boldsymbol{x}) = \theta(x_1)\theta(x_2)\cdots\theta(x_m) \quad \text{for } \boldsymbol{x} = (x_1, \ldots, x_m) \in \{0, 1\}^m$$

factorises according to such an unconnected graph. We can see here that the calculation of every primitive probability depends on one variable only and not on the values taken by any other variable in the system. In terms of an $\boldsymbol{X}$*-compatible staged tree representation for this model, we would draw an event tree with binary florets where all florets which lie along the same level are also in the same stage. This is because every* X_i *is independent of all* $X_1, \ldots, X_{i-1}$ *depicted downstream of the associated vertex,* $i = 2, \ldots, m$. *Thus, this staged tree is stratified and every stage is also a position. The corresponding CEG in Figure 3.7 is an* $\boldsymbol{X}$*-compatible representation of the binary independence model. We denote here by* w_i *the position whose edges are identified with the state space of the random variable* X_i, $i = 1, \ldots, m$, *so the root is in this example denoted* $w_0 = w_1$.

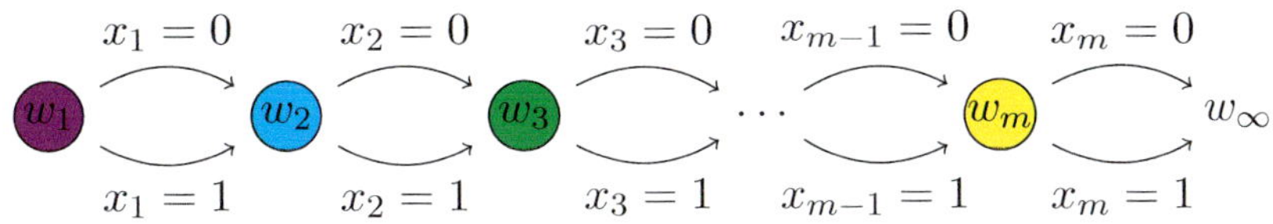

FIGURE 3.7
A CEG representation for the binary independence model from Example 3.12.

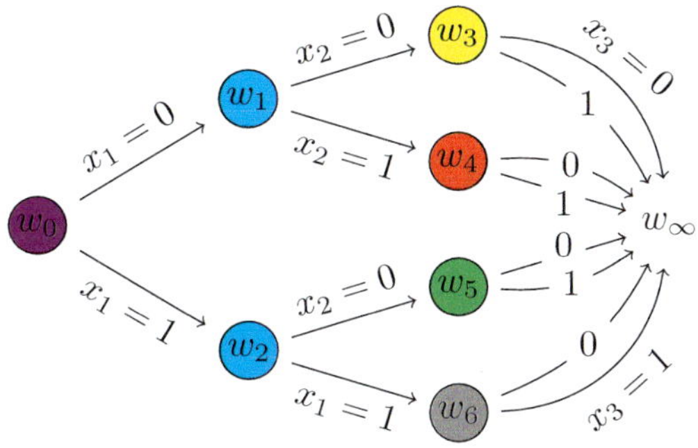

FIGURE 3.8
A CEG representing the collider in Example 3.13.

Example 3.13 (Collider graphs). *Consider again a binary random vector* $\boldsymbol{X} = (X_1, X_2, X_3)$*, and suppose now that* X_1 *and* X_2 *are independent,* $X_1 \perp\!\!\!\perp X_2$*, with no assumptions made on* X_3*. We can represent this model in the* collider *DAG below:*

$$\mathcal{G}: \; X_1 \to X_3 \leftarrow X_2$$

The probability mass function $p_{\boldsymbol{\theta}}(x_1, x_2, x_3) = \theta(x_1)\theta(x_2)\theta(x_1, x_2, x_3)$ *factorises according to* $\mathcal{G}$*, for* $(x_1, x_2, x_3) \in \{0, 1\}^3$*. Just as in the example above, we can draw a staged tree with eight root-to-leaf paths, each of which represents an atom in the state space. This tree has a root-floret depicting the state space of* X_1*, two florets on the first level depicting* $X_2|X_1 = 0$ *and* $X_2|X_1 = 1$ *and four florets on the second level depicting the conditional random variable* $X_3|X_1 = x_1, X_2 = x_2$ *for varying values of* $x_1, x_2 \in \{0, 1\}$*. The explicit construction is again left to the reader.*

In this $\boldsymbol{X}$*-compatible tree, both florets on the first level are in the same stage because* X_1 *and* X_2 *are independent, just like in Example 3.12. However, no vertices on the second level of the tree are identified because* X_3 *depends on both* X_1 *and* X_2*. So the stage on the first level is not a position and the resulting CEG—as depicted in Figure 3.8—will have two vertices in the same stage. This CEG is* $\boldsymbol{X}$*-compatible and stratified.*

Within this section we have emphasised the CEG as a tool which embellishes standard BN models. This is so that the reader can compare our new models with more familiar classes of graphical models. However, as we have discussed throughout Chapter 1, it is critical to be aware that many modelling situations that can be expressed as a CEG simply do not enjoy a natural BN structure. This is especially true when information cannot be meaningfully and fully described in terms of a simple vector of values of a random vector. We continue to demonstrate the extra expressability of the CEG class in Chapters 6 and 7. All of the methods developed in Chapters 4 and 5 and early in Chapter 6 apply in the most general setting.

3.4 Examples of CEG semantics

The class of different CEG models $\boldsymbol{C}(\Omega) = \{\boldsymbol{M}_{\mathcal{C}} \mid \mathcal{C} \text{ is a CEG representing } \Omega\}$ over a given finite set Ω of atomic events is often enormous. Although the general class is extremely rich, in practice we often find that in any given context only particular types of Chain Event Graphs $\mathcal{C}$ will make sense. Obviously not only when choosing a particular representation but also when searching for a 'best fitting' CEG is it important that the resulting model does indeed make sense: so that it lies in a contextually defined subclass $\{\boldsymbol{M}_{\mathcal{C}} \mid \mathcal{C} \text{ is a CEG representing } \Omega \text{ and has certain properties}\} \subsetneq \boldsymbol{C}(\Omega)$.

We list some of the most useful of these classes below.

3.4.1 The saturated CEG

Definition and example

The class of *saturated* CEGs is the class $\boldsymbol{C}_0(\Omega) \subseteq \boldsymbol{C}(\Omega)$ of CEG models on a prespecified set of atoms Ω such that each inner stage of every staged tree representing a model in this class contains precisely one situation and its leaves are all members of the same final stage.

Figure 3.2 depicts a saturated staged tree. See also Examples 4.14 and 4.21 in the next chapter.

Properties

A candidate CEG $\mathcal{C}_0 = \mathcal{C}(\mathcal{T})$ representing a model $\boldsymbol{M}_{\mathcal{C}_0} \in \boldsymbol{C}_0(\Omega)$ has a staged tree $\mathcal{T}$ for which every vertex is the only member of its stage, so no floret probabilities are associated with each other and there are no additional irrelevance statements added to the structure of that graph. In particular, every CEG in this class is identical to its staged tree representation except for the leaves which can be merged into a sink node.

As a consequence, there are no constraints over the underlying probability space over Ω and the associated model always equals the full probability simplex, $\boldsymbol{M}_{\mathcal{C}_0} = \Delta^{\circ}_{\#\Omega-1}$ [43].

Many useful classes of CEGs contain $\boldsymbol{C}_0(\Omega)$ as a special case: see below.

In developing model selection and learning techniques over Chapters 5 and 6 we often start from considering saturated staged trees and CEGs first. For instance, from a domain expert we can first elicit an explanation in terms of a tree graph and only then elicit its stage structure. However, saturated CEGs are most important because of a more technical result. It was shown in [33] that if we were to use Bayes Factor scoring methods as in Section 2.3 and wanted for all scores to be the same whatever the tree representation of these atoms, then with some minor regularity conditions we needed to use independent Dirichlet distributions on the probabilities associated with edges of each floret in a saturated CEG. So if $\boldsymbol{C}_0(\Omega)$ is contained in a larger search class then this gives quite a strong rationale—one that can be justified by demanding comparability of priors over different parts of a model in the larger class—for using prior Dirichlet distributions on the edge probabilities over all the candidate CEGs in that larger class. We present details of this approach in Section 6.1.

Deficiencies

Used on its own, the class of saturated CEGs is nothing new of course: it is completely analogous to a class of models represented by unconstrained probability trees. These standard classes of probability trees are well studied [90]. Furthermore, with no conditional independence relationships expressed in it, the class of saturated CEGs is very unstable to estimate with the usual parameter independence assumptions we might make even for moderately large problems because the dimension of the probability vector over Ω—the number of atoms minus one—is typically an enormous number. This means that saturated trees are arduous to elicit. Furthermore, in model selection the number of atoms is often much greater or at least of the same order of magnitude as the sample size of any available dataset. This means that any good model selection method—one that tends to favour simpler models over more complex ones when there is little evidence in favour of the complex one—would only very rarely choose a model from this class.

Despite these drawbacks many selection methods would begin with such a saturated model and sequentially improve it. As a consequence, this class provides a good benchmark for measuring how much extra value we are getting from associating various floret probabilities with each other in the more elaborate models discussed in this section: see also Section 6.3.

3.4.2 The simple CEG

Definition and example

The class $\boldsymbol{C}_{\text{simple}}(\Omega) \subseteq \boldsymbol{C}(\Omega)$ of *simple* CEG models contains only models represented by CEGs which are such that in the underlying staged tree, all stages are also positions—so CEGs for which every colour appears exactly once.

Figure 3.5b depicts a simple CEG whilst the CEG depicted in Figure 3.4 is not simple.

Properties

The class of simple CEGs is a very useful one for a number of reasons. Of course any CEG in $bsC(\Omega)$ has another CEG lying in $\boldsymbol{C}_{\text{simple}}(\Omega)$ which simply makes less modelling assumptions—we simply forget all the conditional independences that are expressed by vertices in that graph which are stages but not positions. This class can be identified with the class probability decision graphs we discuss in the next section, albeit with a rather different semantic: see Section 3.5. Again also this class contains the saturated class $\boldsymbol{C}_0(\Omega) \subseteq \boldsymbol{C}_{\text{simple}}(\Omega)$ as a special case.

The class of simple CEGs is entirely graphical. It is therefore very simple to develop fast propagation algorithms around it: see Chapter 5 and [108]. There is also now a d-separation theorem for this class available which is analogous to the d-separation theorem for Bayesian networks and reads *all* conditional independences: see Section 5.2 and [106].

Deficiencies

The class of simple CEGs is quite restricted and does not even contain the class of BNs. So although it is good for representing implicit conditional independences in highly asymmetric models it often does not contain some models we might like to entertain.

3.4.3 The square-free CEG

Definition and example

The class $\boldsymbol{C}_{\text{square-free}}(\Omega) \subseteq \boldsymbol{C}(\Omega)$ of *square-free* CEG models is represented only by graphs for which no two situations that lie on the same root-to-leaf/sink path can also be members of the same stage.

All examples we have seen so far are examples of square-free CEGs. Examples of repeated experiments such as Example 4.13 in the next chapter are not square-free.

Properties

All the classes mentioned above are subclasses of this class. We call CEG representing models in $\boldsymbol{C}_{\text{square-free}}(\Omega)$ square-free because the corresponding monomial parametrisation (3.1)—and hence the polynomial we define in Section 4.2—is square-free.

Square-free CEGs have turned out to be extremely important. For instance, equivalence classes of square-free CEGs representing the same statistical model can be determined in a straightforward way [43, 47]. Furthermore, causal effects associated with this model correspond to simple differentiation operations on polynomials: see Chapter 8 and [46]. If a CEG lies outside this class then we need more structure to be specified before a sample distribution can be unambiguously identified.

The class of square-free CEGs naturally contains the class of all Bayesian networks which are not dynamic.

Deficiencies

This is a very large class of models and so difficult to fully search across except when the number of atoms of the associated space is small. However not all models are square-free. In particular, many useful classes of dynamic models—where structure is naturally repeated along root-to-sink paths—are members of this class.

3.5 Some related structures

In the past, there have been various different approaches to use probability trees to accurately describe discrete modelling problems. In most of these, the authors have used probability trees exclusively as an elegant representation to depict a process of interest rather than as a formal structure. This was also often motivated by applications of decision analysis rather than statistics. None of these alternative uses of probability trees included a formal representation of conditional independence assumptions—as our graphs include colour for that purpose—and all of these have very different semantics, which are not straightforward to translate into well-known frameworks such as the one for Bayesian networks.

Probability trees in causal inference

One of the earliest formalisations of probability trees in mathematical language was achieved by Shafer and mostly aimed at their application in causal inference [90]. Here, the event tree was first used as a vehicle to capture the philosophical idea of causality within a rigorous framework, making it possible to extend a simple tree diagram to answer queries in probabilistic and then in causal inference. Most of the language we use for event trees and CEGs is inspired by this foundational work.

We will outline in detail in Chapter 8 why event trees (and hence CEGs) are particularly expressive in terms of depicted putative causal hypotheses. A few early observations of this point are also presented in Section 4.1 where we interpret an event tree as a picture for a partial order on a set of events.

Probabilistic Decision Graphs

One of the successful attempts to use probability trees on the interface between artificial intelligence and probabilistic inference is given by *probabilistic decision graphs (PDG)* [53, 54]. A PDG is a DAG where vertices represent random variables lying within a prespecified class. There is a directed edge coming out of a vertex for every state the associated variable can assume. An edge exists between two vertices if one is known to be a direct successor of the other. A vector of labels which sum to one but are not necessarily positive is also attached to each vertex.

Conditional independence assumptions can be read from the PDG using certain partitions of the state space. These cannot in general be expressed in a BN. However, unlike for the stratified CEG not all BN models can be represented by a PDG. This can nonetheless be a very useful representational device. For example, the authors cited above have developed an algorithm to efficiently transform a given junction tree into a PDG while preserving its associated distributional assumptions.

Using functions which are heavily based on the structure of the graph, a PDG can also be used to answer simple probabilistic queries such as the computation of marginal probabilities. Just like CEGs and BNs, PDGs can be either directly specified by a domain expert or learned from data.

Probability trees for importance sampling

Probability trees can further be used in conjunction with Monte Carlo methods for fast propagation in Bayesian networks [88]. In this application again inner nodes correspond to prespecified random variables and edges are as above labelled by the probabilities of their respective state spaces. The atomic probabilities associated to the leaves of such a tree can then be used to store potentials, so unnormalised conditional or marginal probabilities.

In stark contrast to our use of event trees, in this application of probability trees the propagation algorithm goes through a pruning step where it removes all vertices associated to random variables which are independent of a variable of interest. So rather than specifying a model using a coloured tree which represents exactly these constraints, the probability trees here will not depict any quantities which are irrelevant to a given computational query.

Coalescent Trees

Coalescent trees have been well-known objects in decision theory for some time [13]. In a coalesced tree, vertices with equal (or isomorphic) future unfoldings are combined in the same way as vertices in the same position can be merged into a single vertex in a CEG. The CEG naturally exploits this idea of coalescence. However, general coalescent trees cannot easily depict conditional independence assumptions.

This approach is in a sense closest to ours and ways that CEG semantics can be used within a decision theoretic framework have recently been devised to enhance earlier work [107].

3.6 Exercises

Exercise 3.1. *Consider Example 3.7. Draw an alternative staged tree describing the same problem but this time taking into account that the probability of carrying an antidote is independent of the actual encounter with the snake. Discuss which representation is more apt to represent different types of queries a modeller might be interested in. Can you think of one of the staged trees as representing a submodel of the other?*

Exercise 3.2. *Draw a CEG for which one of the three properties* stratified, simple *and* square-free *is true and the other two are false. Draw a CEG where two properties are true and the third one is false.*

Exercise 3.3. *Draw* $\boldsymbol{X}$*-compatible staged tree representations for the CEGs in Examples 3.11 and 3.13.*

Exercise 3.4. *Let* $\boldsymbol{X} = (X_1, X_2, X_3, X_4)$ *be a vector of binary random variables for which we assume that the conditional independence assumptions* $X_1 \perp\!\!\!\perp X_4 \mid X_2, X_3$ *and* $X_2 \perp\!\!\!\perp X_3$ *are simultaneously true. Write down the probability mass function and represent this assumption in a DAG. Then draw* $\boldsymbol{X}$*-compatible staged tree and CEG representations for this model. Finally, add the context-specific information* $X_1 \perp\!\!\!\perp X_3 \mid X_2 = 0$ *first in the staged tree and then in the CEG.*

Exercise 3.5. *Check that Proposition 3.10 is true for Examples 3.11 to 3.13.*

Exercise 3.6. *Find a CEG which is stratified but not* $\boldsymbol{X}$*-compatible, so show that the class of CEG models is strictly larger than that of context-specific BNs.*

4

Reasoning with a CEG

In this chapter we give a careful formal development of CEG models. In particular, we will answer two major questions which form the pillars of a successful use of CEGs in statistical inference:

Q1 How can we embed a given problem description into a Chain Event Graph and, conversely, how can we read underlying modelling assumptions from a given CEG?

Q2 When do two CEGs represent the same model and so what can we legitimately draw out of a particular CEG representation?

In answering the first query we will develop notions for CEGs which are analogous to the ideas of d-separation and local and global independence in BNs. This will be presented in Section 4.1 below. The second question then addresses the concept of statistical equivalence in staged trees and CEGs, a notion analogous to distribution-equivalence, and hence Markov-equivalence, in BNs. We address this issue in Section 4.2. Both of these concepts were introduced for BNs in Section 2.3 and have been thoroughly analysed in the standard literature on graphical models. It is hence only natural to develop analogous notions for CEGs.

The methodological development we need in order to answer Q1 and Q2 is based on a number of rather technical results. These results are presented in greater depth in the publications that we cite throughout but are given here in a modified and extended form. In particular, in this book we focus first and foremost on illustrating the key ideas through a number of new examples.

4.1 Encoding qualitative belief structures with CEGs

This section will provide three key inferential tools for the reader. First, we show how to interpret a CEG from a formal probabilistic perspective, especially showing how this graph can express a space of events in a way that is sympathetic to hypotheses about *how* situations might happen. This leads us, in technical jargon, to study the sigma algebra of events that is generated by atoms which are root-to-sink paths of a CEG. This development is particular to the CEG, rather than the BN, in the sense that this sigma algebra is depicted *explicitly* in the graph. Second, we determine which types of events can be depicted unambiguously within a CEG. These events will be called 'intrinsic' and we will illustrate why they are especially important to study. Third, we will relate the colouring and the graph of a CEG to distributional assumptions on the underlying statistical model and, in particular, to their implied conditional independence assumptions. This construction fleshes out in a consistent way the embedded structural hypotheses explicit in a CEG into a full family of probability distributions. We will then be able to use the concepts developed here in the next two chapters to use standard statistical methods, especially Bayesian ones, to estimate, select and analyse a class of CEG models.

An early presentation of many of these results was given in [98, 106].

4.1.1 Vertex- and edge-centred events

In Chapter 3 we have illustrated how useful the probability tree and the CEG can be in the depiction of a given problem. We will now prepare the framework for showing that these graphs are equally useful from a probabilistic perspective. In particular, we first show how to draw out an explanation of a probability tree or CEG in natural language. Two types of events play a special role in this development.

Thinking probabilistically, for any CEG $\mathcal{C}$, the set $\Lambda(\mathcal{C})$ of root-to-sink paths is the set of all atomic events of an underlying probability space Ω. As a consequence, every outcome in the event space—so every element in some sigma algebra of Ω—of a model $\boldsymbol{M}_{\mathcal{C}} \in \boldsymbol{C}(\Omega)$ represented by this CEG can be represented as a subset of this set of atoms. In more technical terms, every event is an element of the *path-sigma algebra* we denote $\sigma(\mathcal{C})$. In this finite setting, $\sigma(\mathcal{C})$ is simply the power set of the set of root-to-sink paths. Here we use a terminology and notation completely analogous to the vocabulary developed for staged trees in the previous chapter. This will enable us to toggle between these two classes seamlessly.

Within the path-sigma algebra of a staged tree or CEG, vertex- or edge-centred events—that is events consisting of all root-to-sink paths passing through a fixed vertex or edge as introduced in Section 3.1—are of key importance in communicating such a graph to a non-statistician. This is because these events are often very straightforward to interpret whenever there is a supporting story about the potential unfoldings of the future as told by that staged tree or CEG. We will see numerous examples of this below.

Every atomic event depicted in a staged tree is in particular the vertex-centred event centred at its final leaf. So for trees, root-to-leaf paths are in one-to-one correspondence with leaf-centred events. In a CEG where all leaves are merged into a common sink-node, this identification is not possible. However, in CEGs we can alternatively think of a root-to-sink path as the intersection of all edge-centred events along that sequence of edges. As a result, vertex- and edge-centred events span the path-sigma algebras of our graphs.

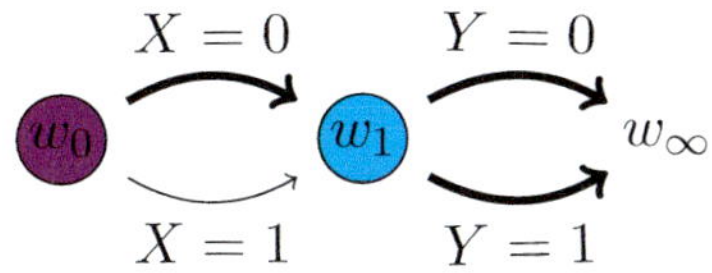

FIGURE 4.1
A CEG representing the independence model in Examples 4.2 and 4.4. The thick depicted event $X = 0$ is an edge-centred, intrinsic event.

Proposition 4.1 (The path-sigma algebra of events). *Every event depicted by a staged tree or CEG can be written as the finite intersection and union of vertex- and edge-centred events.*

This result enables us to focus on very simple objects when reading a CEG. In particular, in $\boldsymbol{X}$-compatible staged trees and CEGs, vertex- and edge-centred events can be easily expressed in terms of the variables $\boldsymbol{X}$. We illustrate this in a simple example below.

Example 4.2 (Edge- and vertex-centred events). *Let $X \perp\!\!\!\perp Y$ be two independent binary random variables. The CEG depicted in Figure 4.1 is (X, Y)-compatible and depicts this independence assumption. Let $\lambda_{(i,j)}$ denote the root-to-sink path in the CEG representing the atomic event $X = i \wedge Y = j$ for $i, j = 0, 1$. For instance, $\lambda_{(0,0)}$ denotes the root-to-sink path corresponding to the atom $(x, y) = (0, 0)$ in the state space $\mathbb{X} \times \mathbb{Y} = \{0, 1\}^2$ of X and Y.*

Now observe that the event $X = 0$ can be written as the union of the two atomic events $X = 0 \wedge Y = 0$ and $X = 0 \wedge Y = 1$. These are represented by the two root-to-sink paths $\lambda_{(0,0)}$ and $\lambda_{(0,1)}$, thick depicted in the figure above. Hence, the union of these two paths is an edge-centred event $\Lambda(e) \in \sigma(\mathcal{C})$ in the CEG, centred on the upper edge e emanating from the root w_0—so precisely the edge labelled $X = 0$. We can see here that edge-centred events in this CEG naturally correspond to marginal events on the underlying random variables.

The vertex-centred event $\Lambda(w_1)$ collects all units arriving at the position w_1. These are all units for which either $X = 0$ or $X = 1$ is true. With no assumptions on the random variable Y, this event thus contains all possible outcomes of the model. In fact, $\Lambda(w_1) = \Lambda(\mathcal{C})$ is the set of all root-to-sink paths of the CEG.

Whenever a position w in a CEG is upstream of another position w', we can order their vertex-centred events in the same way and say that $\Lambda(w)$ is upstream of $\Lambda(w')$, in symbols $\Lambda(w) \prec \Lambda(w')$. For the scope of this chapter, we will merely interpret an event being upstream of the other as saying that in the setting presented by the CEG at hand, this one event *can* happen before the other. This order can then be interpreted in a logical or chronological sense. For instance in the forensic case example from Section 1.2, the staged tree we used to represent this setting was clearly set up following a chronological order of events happening. Similarly, in the cerebral palsy example in Section 1.3

situations unfold in a plausible but logically entailed order: see also Examples 4.10 and 4.11 below. In Section 8.3, we will outline conditions under which this ordering can also be given a putative causal interpretation. In close analogy with BNs, when drawing a CEG elicited from a client it is nearly always good practice to follow as closely as is possible any ordering of events given by the domain expert. We explain in this section how this can be done.

If we follow a unit consecutively passing from the root through a number of vertices, then the graph representing this unfolding tells us exactly how these events are nested within the path-sigma algebra. This is because when passing along one fixed edge in a floret, all other edges in that floret—and all their future unfoldings—become impossible. In this way, a CEG elegantly depicts all events which are *counterfactual* to a given event which we know has happened within a particular story. At the same time, all events depicted downstream of a fixed edge represent possible future unfoldings for a unit going along that edge, and so all of their implications are valid for that unit.

In simple words, a CEG can be interpreted as a picture of a *partially ordered* set of vertex-centred events which can be formally linked to the underlying description from which these events are drawn.

Example 4.3 (Partial orders). *Recall the snakebite example, Example 3.7, from the previous chapter with CEG representation given in Figure 3.5b. The initial situation 'contact with the snake' is here represented as the root-vertex w_0. The vertex-centred event $\Lambda(w_0)$ then depicts all possible unfoldings following from that contact, or our full space of events.*

The downstream event $\Lambda(w_1)$ can be interpreted as 'bitten by a snake'. This vertex-centred event now requires contact with the snake as a prior event and contains all amenable downstream unfoldings, here cure and survival, as subsets. In the alternative staged tree representation of this setting given in Figure 3.5a, $\Lambda(w_1) = \Lambda(v_1) = \Lambda(v_7) \cup \Lambda(v_9) \cup \Lambda(v_9) \cup \Lambda(v_{10})$ is again a vertex-centred event and can be alternatively expressed as the union of four root-to-leaf paths of the form (snakebite, cure, survival) *where the variables cure and survival take the values* (no) antidote *and* dead *or* alive, *respectively. Passing further downstream, the event $\Lambda(w_2)$ depicts all units in an 'endangered' situation, namely those who share the common path from w_0 to w_2 of having been bitten and not having applied an antidote.*

In contrast, the alternative unfolding from the root w_0 to w_3 collects all 'safe' units in the event $\Lambda(w_3)$ complementary to $\Lambda(w_2)$. These are of the form (no bite, survival) *with survival-variable as above. Note here that $\Lambda(w_3)$ is again a vertex-centred event in the CEG but it is the union of vertex-centred events in the corresponding staged tree: $\Lambda(w_3) = \Lambda(v_2) \cup \Lambda(v_3)$ where v_2 and v_3 are in the same stage in the tree above.*

So the CEG of this example orders the events in its path-sigma algebra as follows: $\Lambda(w_0)$ lies upstream of $\Lambda(w_1)$ which in turn lies upstream of $\Lambda(w_2)$. Furthermore $\Lambda(w_3)$ lies downstream of $\Lambda(w_0)$. In other words, snakebite occurs before a potential cure which in turn occurs before a measure of survival. Using *the tools from Section 4.2 below, we can show that this ordering of events is the only one amenable to this model.*

We can see in the example above that a CEG can be used to transparently talk about the relationships between its depicted events, and this even more straightforwardly than in the corresponding staged tree.

In a more complex applied setting, a specification of a model as above might have been elicited in a discussion between the modeller and a domain expert. In this case, we would always start with an explanation of the model in natural language: encounter with the snake first and then subsequent events which are only made possible from certain prior unfoldings, like units being endangered only if they have been bitten. This explanation follows precisely the partial order we then depict in the graph. A model that has been elicited in this way can then be verified using observational data or by discussing all of its implications in terms of the nested partial order of events (or by determining all statistically equivalent representations) with the expert. Exercise 4.1 below analyses which alternative depictions and hence which alternative models could have been used for this description of the problem.

4.1.2 Intrinsic events

When drawing a CEG from a given discrete problem description, there are a number of technicalities which have to be taken into account. In particular, because there are often multiple edges coming into and out of each position, the graph might provide an ambiguous picture of a set of events of interest.

Observe first that every event $\Lambda \in \sigma(\mathcal{C})$ depicted in a CEG $\mathcal{C}$ induces a *subgraph* $\mathcal{C}_\Lambda \subseteq \mathcal{C}$ of the original CEG. The root-to-sink paths of this subgraph are precisely the elements in Λ, so its edges and vertices are given by the edges and vertices of the root-to-sink paths in this event. This subgraph is in general not a CEG. For instance, in Figure 4.1 the event $X = 0$ induces a thick-depicted subgraph which is not a CEG itself. This is because it contains only a single unfolding from the root, and thus has no underlying event tree.

We will henceforth say that an event is *intrinsic* to a CEG if its subgraph representation is one-to-one with a formal description of that event in terms of a collection of atoms, so if and only if $\Lambda \mapsto \mathcal{C}_\Lambda$ is an invertible map. We illustrate below how non-intrinsic events can cause issues in CEG representations, depicting subgraphs of CEGs which contain more root-to-sink paths than those needed to draw these graphs.

Translated into graphical terms, we further define an *intersection* of subgraphs as the intersection of the events which induced those graphs, rather than as the graphical intersection of their vertex and edge sets. So let $\mathcal{C}_\Lambda \cap \mathcal{C}_\Gamma = \mathcal{C}_{\Lambda\cap\Gamma}$ for any events $\Lambda, \Gamma \in \sigma(\mathcal{C})$ in the path sigma algebra. In the same fashion, the *union* of these graphs is here defined as the union of their events, $\mathcal{C}_\Lambda \cup \mathcal{C}_\Gamma = \mathcal{C}_{\Lambda\cup\Gamma}$. Some simple exercises on these notions are given at the end of this chapter: see for instance Exercise 4.2.

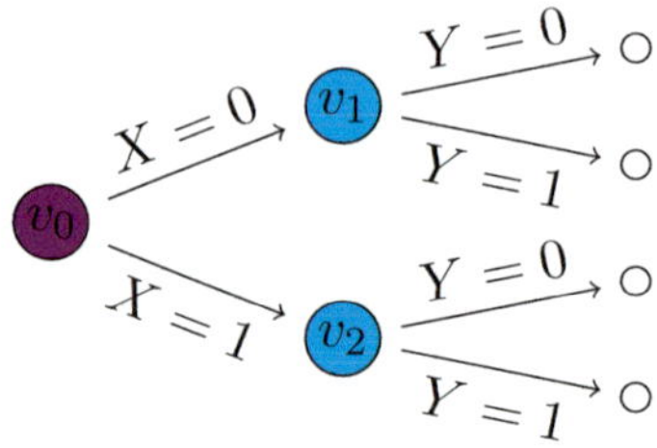

FIGURE 4.2
A staged tree representing the independence model discussed in Example 4.4 and with corresponding CEG given in Figure 4.1. In this graph, the event $\{(0,0),(1,1)\}$ is intrinsic.

Example 4.4 (Intrinsic and non-intrinsic events). *Return to the CEG from Example 4.2 and Figure 4.1.*

We have already seen that the event $X = 0$ can be represented by the union of two root-to-sink paths $\Lambda = \{\lambda_{(0,0)}, \lambda_{(0,1)}\}$. The corresponding subgraph $\mathcal{C}_{\{\lambda_{(0,0)},\lambda_{(0,1)}\}}$ was thick depicted in Figure 4.1. This subgraph contains precisely the two root-to-sink paths $\lambda_{(0,0)}$ and $\lambda_{(0,1)}$ and no other. The event $\{0\} \times \{0,1\} \subseteq \{0,1\}^2$ which gives rise to this subgraph is thus an intrinsic event in the CEG.

However, for instance the event $\{(0,0),(1,1)\} \subseteq \{0,1\}^2$ is not intrinsic. This is because this event is represented as the union of two root-to-sink paths $\{\lambda_{(0,0)}, \lambda_{(1,1)}\}$ for which corresponding subgraph $\mathcal{C}_{\{\lambda_{(0,0)},\lambda_{(1,1)}\}}$ has the same vertex and edge set as the whole CEG $\mathcal{C}$. Thus graphically, these two CEGs are the same. But clearly $\{\lambda_{(0,0)}, \lambda_{(1,1)}\}$ is not equal to the whole set of atomic events $\{\lambda_{(0,0)}, \lambda_{(0,1)}, \lambda_{(1,0)}, \lambda_{(1,1)}\}$. So in this case, the map which defines a subgraph $\{\lambda_{(0,0)}\ \lambda_{(1,1)}\} \mapsto \mathcal{C}_{\{\lambda_{(0,0)},\lambda_{(1,1)}\}}$ is not invertible. As a consequence, the event depicted here is not intrinsic.

Typically, a CEG will have many non-intrinsic events. However, these are usually events that have no role in the story underlying the CEG. This might be for example because they are given by the union of atoms which are not connected in the narrative. We will further discuss this point below.

We can prove a number of interesting properties of intrinsic events. For instance, intersections of intrinsic events are always intrinsic—we leave this as Exercise 4.3 to the reader—but unions of intrinsic events are not necessarily intrinsic. An example of this has been provided above. Vertex- and edge-centred events are by construction always intrinsic, so are atomic events and so is the whole space of events in a given CEG.

Because in event trees every vertex has a unique entering edge, for staged trees every event is intrinsic. Furthermore, because the sets of events in a staged tree are in one-to-one correspondence with the sets of events in a CEG, and because both are representations of the same model, issues in communicating CEGs which are caused by non-intrinsic events can always be avoided by choosing to represent a certain setting by a staged tree rather than a CEG. See for instance Figure 4.2 which solves this issue for Example 4.4.

4.1.3 Conditioning in CEGs

Having covered the more graph-theoretical aspects of specifying events in a Chain Event Graph above, we can now turn to using these results in inference. In this development, we will use the terminology *subCEG* for a CEG whose edge and vertex sets are subsets of a given bigger CEG. Note that in contrast to the subgraphs we analysed in the previous section, subCEGs are not necessarily induced by a set of events. However, subCEGs can be transparently used as graphical representations of conditioning operations and are in that sense subgraphs induced by conditional events—often conditional on arriving at a certain vertex or passing along a certain edge. If this is the case then it is central to know how primitive probabilities translate from the bigger graph into primitive probabilities in a subCEG. We will present details of this below, following the initial development in [106].

First note that the probability of a vertex-centred event in a given staged tree equals precisely the probability of a unit arriving at that vertex. This is because all root-to-leaf paths take the same subpath from the root to that vertex and because the induced subtree rooted at that vertex has atomic probabilities which sum to unity. So for all events $\Lambda(v) \in \sigma(\mathcal{T})$ centred at vertices $v \in V$, represented in an event tree $\mathcal{T} = (V, E)$, we find that

$$\begin{aligned} P_{\boldsymbol{\theta}}(\Lambda(v)) &= \sum_{\lambda \in \Lambda(v)} p_{\boldsymbol{\theta},\mathcal{T}}(\lambda) = \prod_{e \in E(v_0 \to v)} \theta(e) \Big(\sum_{\lambda' \in \Lambda(\mathcal{T}(v))} p_{\boldsymbol{\theta},\mathcal{T}(v)}(\lambda') \Big) \\ &= \prod_{e \in E(v_0 \to v)} \theta(e) \end{aligned} \tag{4.1}$$

where $E(v_0 \to v) \subseteq E$ denotes the edge set of the subpath from the root v_0 to the event's central vertex v, $P_{\boldsymbol{\theta}}$ denotes a probability measure associated to the tree and $\mathcal{T}(v)$ denotes the subtree of $\mathcal{T}$ which is rooted at $v \in V$ and whose root-to-leaf paths are precisely the v-to-leaf paths in the bigger tree: see Section 3.1.

The result above is equally true for the CEG: the probability of arriving at a position-centred event is the probability of arriving at that position. So here in the final monomial in (4.1) we would sum over all vertices in the same position as the given central vertex:

$$P_{\boldsymbol{\theta}}(\Lambda(w)) = \sum_{v \in w} \prod_{e \in E(v_0 \to v)} \theta(e) \tag{4.2}$$

for any position w in a CEG and using notation for the underlying staged tree as in (4.1).

Second, observe that the edge labels of a staged tree or CEG are indeed conditional probabilities. In fact, for every edge $e = (v, v') \in E$ in a staged tree, the primitive probability $\theta(e)$ is the probability of passing on to v' given that a unit has arrived at the situation v:

$$\theta(v, v') = P_{\boldsymbol{\theta}}(\Lambda(v') \mid \Lambda(v)) = \frac{P_{\boldsymbol{\theta}}(\Lambda(v'))}{P_{\boldsymbol{\theta}}(\Lambda(v))} \tag{4.3}$$

in the notation above. Here, one vertex-centred event is contained in the other $\Lambda(v') \subseteq \Lambda(v)$. This is because their central vertices are connected by an edge and so all root-to-leaf paths passing through v' must have passed through the unique parent v.

This statement can simply be proven by writing (4.3) in terms of (4.1) above and cancelling the terms which appear in both numerator and denominator of the fraction [43].

Example 4.5 (Conditional probabilities)**.** *We are now interested in calculating the probability of death given that a unit has been bitten, and of survival given that it is safe in Examples 3.7 and 4.4.*

First, using the formula (4.1) above, the probability of being bitten by a venomous snake equals the product of all edge labels from w_0 to w_1, so simply $P_{\boldsymbol{\theta}}(\Lambda(w_1)) = \theta_{\text{bite}}$ in the obvious notation. Note that indeed the probabilities of all downstream unfoldings sum to one in this graph. Similarly, the probability of a unit in the system being safe equals the probability of reaching position w_3, so $P_{\boldsymbol{\theta}}(\Lambda(w_3)) = \theta_{\text{bite}}\theta_{\text{antidote}} + \theta_{\neg\text{bite}}$ where again the future unfoldings from w_3 sum to unity. Then the desired conditional probabilities can be calculated as

$$\begin{aligned} P_{\boldsymbol{\theta}}(\text{dead} \mid \text{snakebite}) &= \frac{1}{\theta_{\text{bite}}}\big(\theta_{\text{bite}}\theta_{\neg\text{antidote}}\theta_{\text{dead}} + \theta_{\text{bite}}\theta_{\text{antidote}}\theta_{\text{dead}}\big) = \theta_{\text{dead}} \\ P_{\boldsymbol{\theta}}(\text{alive} \mid \text{safe}) &= \frac{1}{\theta_{\text{bite}}\theta_{\text{antidote}} + \theta_{\neg\text{bite}}}\big(\theta_{\text{bite}}\theta_{\text{antidote}}\theta_{\text{alive}} + \theta_{\neg\text{bite}}\theta_{\text{alive}}\big) \\ &= \theta_{\text{alive}}. \end{aligned}$$

In simple words, given a unit has been bitten, its probability of death is simply given by the primitive probability on the edge labelled dead *downstream of the position* snakebite. *Similarly, the probability of being alive given that a unit is safe is simply the primitive probability of the edge labelled* alive.

Using the usual definition of conditional probability, when conditioning on a vertex-centred event in a CEG we now simply divide every primitive probability in the graph by the quantity calculated in (4.1) above. For edges upstream of that vertex, all probabilities are thus set to one because the fraction cancels out—the corresponding edge-centred events are assumed to have happened. For edges downstream of that vertex, the primitive probabilities will be renormalised such that atomic probabilities sum to unity. This translates into the following result.

Proposition 4.6 (Conditioning). *Let $(\mathcal{C}(\mathcal{T}), \boldsymbol{\theta}_{\mathcal{T}})$ be a CEG with associated probability measure $P_{\boldsymbol{\theta}}$, and let $\Lambda \in \sigma(\mathcal{C})$ be an intrinsic event. Then in the subgraph $\mathcal{C}_{\Lambda}$, all primitive probabilities are of the form*

$$\theta(e)|_{\mathcal{C}_{\Lambda}} = \frac{P_{\boldsymbol{\theta}}(\Lambda \mid \Lambda(e))}{P_{\boldsymbol{\theta}}(\Lambda \mid \Lambda(w))}\theta(e) \tag{4.4}$$

where w denotes the head of the edge $e = (w, w')$.

A proof of this proposition can be found in [106]. In a symbolic framework, conditional probabilities as above can be easily calculated by differentiating a polynomial in the edge probabilities associated to a CEG [45, 67].

The simple result of Proposition 4.6 was shown to be extremely useful [105]. Of course whilst much more powerful than its BN analogous, the CEG machinery we build here cannot answer all queries. In particular, the stage structure in a CEG might be violated by conditioning unless the CEG is simple, so unless all vertices in the same stage are also in the same position: see Section 3.4. For non-intrinsic events, these issues get worse and the resulting 'conditioned CEGs' are sometimes neither CEGs nor subgraphs of CEGs. A more technical presentation of this result, including a number of illustrations, has been given in [106].

It is extremely important to note that the standard structural analysis of BNs *only* entails the study of what might happen conditional on a set of intrinsic events in a corresponding CEG. So the graphical study of intrinsic events in a CEG *contains* all the queries we might be able to answer when studying a DAG instead. To see this consider a vector $\boldsymbol{X} = (X_1, \ldots, X_m)$ of random variables represented by a DAG. Then any conditioning operation of the type $X_i \mid X_j = x_j$ for some $i, j \in \{1, \ldots, m\}$ in a corresponding $\boldsymbol{X}$-compatible CEG is simply conditioning on the intrinsic event centred on the edge labelled $X_j = x_j$. Compare also Example 4.2.

In the subsequent section we will define random variables associated with staged trees and CEGs in order to further refine this point. The results of this and the previous section are then central to the propagation algorithm for CEGs we present in Section 5.2 where the conditioning operation from Proposition 4.6 will be translated into simple local calculations on the graph in order to calculate these quantities fast.

4.1.4 Vertex-random variables, cuts and independence

We have seen in Section 3.3 how to embed a prespecified set of discrete problem variables into an $\boldsymbol{X}$-compatible CEG. In this section, we will take the contrary approach: given a CEG, we will read new random variables from the graph and analyse the conditional independence structures we can infer from the colouring of the CEG. For $\boldsymbol{X}$-compatible staged trees, these approaches are inverse to each other. For more general staged trees, we will be able to construct functions of random variables which could not be specified from alternative graphical models such as BNs.

Let hence $\mathcal{C} = (W, E)$ be a CEG. For each position $w \in W$, we define two *incident* random variables $I_w : \Lambda(\mathcal{C}) \to \{0, 1\}$, and $Y_w : \Lambda(\mathcal{C}) \to \mathbb{R}$ as follows:

$$I_w(\lambda) = \begin{cases} 1 & \text{if } \lambda \in \Lambda(w) \\ 0 & \text{otherwise} \end{cases} \quad \text{and} \quad Y_w(\lambda) = \begin{cases} y(w') & \text{if } (w, w') \in E(\lambda) \\ 0 & \text{otherwise.} \end{cases} \tag{4.5}$$

The first random variable above is an indicator random variable of whether or not a root-to-sink path passed through a specific position. The latter is a random variable which takes a non-zero value only when a root-to-sink path passes through the edge emanating from that position which is labelled by that value.

Formally, we first note that incident random variables are measurable with respect to the set of root-to-sink paths of a CEG and its associated probability measure. Thus, they enable us to define local *vertex-random variables* X_w as conditional random variables $Y_w \mid I_w = 1$ for $w \in W$. These are now measurable with respect to the subtree (or subCEG) rooted at w.

Thus for every CEG, we can specify a collection of random variables, one for each position in the graph. The state space of these random variables is represented by the edges emanating from that position. By abuse of notation, we can hence write $X_w = w'$ for the vertex-random variable associated to the position w 'taking the value' w'. This is shorthand for $Y_w = y(w') \mid I_w = 1$ in the notation above.

Primitive probabilities in a CEG are thus entries in the conditional probability tables associated to their incident and vertex-random variable. In particular, for every edge $e = (w, w')$, we have

$$\theta(w, w') = P_{\boldsymbol{\theta}}(X_w = w') = P_{\boldsymbol{\theta}}(Y_w = y(w') \mid I_w = 1). \tag{4.6}$$

As a consequence, in staged trees and CEGs two vertices are in the same stage if and only if their vertex-random variables have the same distribution.

Example 4.7 (Vertex-random variables in previous examples)**.** *In the binary independence model from Examples 4.2, 4.4 and 4.8, we can specify vertex random variables $X_{w_0} = X$ and $X_{w_1} = Y$. These are precisely the two random variables which induced this (X, Y)-compatible CEG.*

In the snakebite model from Examples 4.3 and 4.5, we define vertex-random variables as follows. X_{w_0} can be interpreted as the random variable 'snakebite' taking values $x(w_1)$ and $x(w_3)$ which quantify the outcomes yes *and* no*, respectively. X_{w_1} is the variable 'antidote', and X_{w_2} and X_{w_4} are two random variables measuring 'survival', conditional on two different upstream unfoldings: one having the history* safe *and the other* endangered.

We can now use incident- and vertex-random variables in a CEG to read various implicit conditional independence assumptions from the graph. This can be done in a straightforward manner for both the staged tree and the CEG. Specifying these assumptions between random variables rather than between events in the path-sigma algebra gives us an analogoue to the separation theorem for BNs: see Proposition 2.4. Because the concept of conditional independence between these variables is deeply interlinked with the exact location of their respective positions within a CEG, we need some additional terminology before we can state the main result.

Let $\mathcal{T} = (V, E)$ denote the graph of a staged tree. We say that a set of vertices $W \subseteq V$ is a *fine cut* if the disjoint union of events centred on these vertices equals the whole set of root-to-leaf paths, so if none of the vertices within W is up- or downstream of another and $\bigcup_{w \in W} \Lambda(w) = \Lambda(\mathcal{T})$. We will also call the set of vertices in the corresponding CEG a fine cut.

We will further call a set of stages $W' \subseteq U_{\mathcal{T}}$ in the tree a *cut* if the set $\{v \in u \mid u \in W'\}$ is a fine cut. Just as above, the term cut is then also used for the corresponding set of vertices in the alternative CEG representation.

A random variable X_W is called a *cut-variable* if W is a cut or a fine cut and X_W is measurable with respect to the probability space defined by the underlying staged tree or CEG. So in simple words, a cut-variable is an indicator variable for which vertex in a cut a root-to-leaf/sink path has passed through. We further demand that if W is a cut then X_W will take the same value for all vertices in the same stage.

We can then denote by $Y_{\prec W} = (Y_w \mid w \text{ upstream of } W)$ the vector of incident-random variables whose vertices are located upstream of W and by $Y_{W \preceq} = (Y_{w'} \mid w' \text{ downstream of } W)$ the vector of incident-random variables whose vertices are located downstream of W or are elements of W. Then in this notation:

Theorem 2 (Cut-variables and independence). *Let $\mathcal{C} = (W, E)$ be a CEG and let $W' \subseteq W$ be a set of vertices. Then for any cut-variable $X_{W'}$, we find:*

1. *If W' is a fine cut then $Y_{\prec W'} \perp\!\!\!\perp Y_{W' \preceq} \mid X_{W'}$.*
2. *If W' is a cut then $Y_{\prec W'} \perp\!\!\!\perp Y_{W'} \mid X_{W'}$.*

A proof can be found in [98, 106].

The results above can look a bit ominous at first but we find that they are actually very transparent in the examples we present for illustration below. In particular, Theorem 2 is directly analogous to what is known as the directed Markov property in DAGs, where descendants are independent of ancestors given their parents: see Section 2.3. So statements of the type 'the future is independent of the past given the present' can be easily read from a CEG.

A simple corollary of Theorem 2 is that if a cut or fine cut consists of one single vertex that every unit in the graph has to pass through then all unfoldings after that vertex are independent of whatever events happened before that vertex.

Example 4.8 (Independence). *We return for a third time to the binary independence model represented by the CEG in Figure 4.1. Here, the position*

w_1 is a cut and a fine cut. By Theorem 2, the associated incidence random variables $Y_{w_0} \perp\!\!\!\perp Y_{w_1}$ are independent. This implies simply that the random variables X and Y are independent.

We could have alternatively obtained this result as follows. We label the edges of this graph by $\boldsymbol{\theta}_{w_0} = (\theta_0, \theta_1)$ and $\boldsymbol{\theta}_{w_1} = (\tau_0, \tau_1)$. Then for any $i, j = 0, 1$, we can calculate the marginal probabilities

$$\begin{aligned} P_{\boldsymbol{\theta}}(X = i) &= P_{\boldsymbol{\theta}}(\{\lambda_{(i,j)}\} \cup \{\lambda_{(i,\neg j)}\}) = \theta_i \tau_j + \theta_i \tau_{\neg j} \\ P_{\boldsymbol{\theta}}(Y = j) &= P_{\boldsymbol{\theta}}(\{\lambda_{(i,j)}\} \cup \{\lambda_{(\neg i,j)}\}) = \theta_i \tau_j + \theta_{\neg i} \tau_j \end{aligned} \tag{4.7}$$

where $\boldsymbol{\theta} = (\boldsymbol{\theta}_{w_0}, \boldsymbol{\theta}_{w_1})$ and $\neg i \neq i$ and $\neg j \neq j$ also take the values 0 or 1. This yields that

$$\begin{aligned} P_{\boldsymbol{\theta}}(X = i) P_{\boldsymbol{\theta}}(X = j) &= \theta_i (\tau_j + \tau_{\neg j}) \tau_j (\theta_i + \theta_{\neg i}) = \theta_i \tau_j \\ &= P_{\boldsymbol{\theta}}(X = i, Y = j) \end{aligned} \tag{4.8}$$

thanks to the local sum-to-1 conditions on florets. As a result, X and Y are independent binary random variables.

The validity of Theorem 2 can be proven using calculations analogous to the ones presented in the example above. Note that of course Example 4.8 can be generalised to the case of m independent random variables as in Example 3.12. Here we can both read the same independence statements from a CEG as we could from a corresponding BN and in addition have an elegant and transparent depiction of the sample space of the variables in the system—a feature not shared by the DAG representation.

Example 4.9 (Context-specific BNs). *Let X, Y, Z be binary random variables with an associated probability measure P and with conditional independence assumptions as depicted in the (X, Y, Z)-compatible staged tree in Figure 4.3a.*

This tree has two trivial stages $u_0 = \{v_0\}$ coloured violet and $u_1 = \{v_1\}$ coloured red and three non-trivial stages $u_2 = \{v_2, v_3\}$ coloured blue, $u_3 = \{v_4, v_5\}$ coloured yellow and $u_4 = \{v_6, v_7, v_8, v_9\}$ coloured green. These stages capture precisely the assumptions that

$$P(Y = i \mid X = 1) = P(Y = i \mid X = 2) \tag{4.9a}$$

$$P(Z = j \mid X = 0, Y = 1) = P(Z = j \mid X = 0, Y = 0) \tag{4.9b}$$

$$P(Z = j \mid X = 1, Y = i) = P(Z = j \mid X = 2, Y = i) \tag{4.9c}$$

where $i, j = 0, 1$.

Now how do we read the identifications of (4.9) from the corresponding CEG in Figure 4.3b using Theorem 2? Here, we have five vertex-random variables X_{w_i}, $i = 1, 2, 3, 4, 5$, where w_0 is the root, $w_1 = u_1$ is the trivial position, $w_2 = u_2$ is the position that is equal to the blue-coloured stage in the tree, $w_3 = u_3$ is the position that is equal to the yellow-coloured stage in the tree

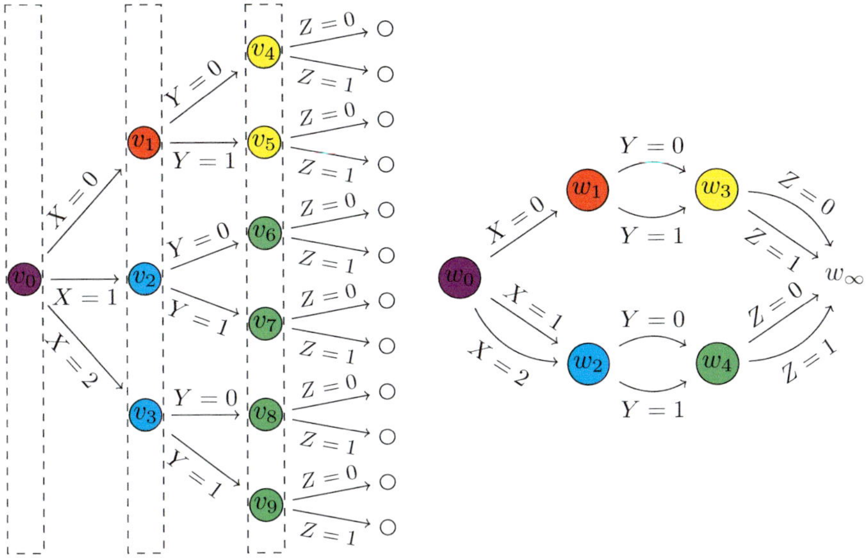

(a) A staged tree. **(b)** A corresponding CEG.

FIGURE 4.3
Two graphical representations for the context-specific BN in Example 4.9 from which we can read the context-specific information directly.

and $w_4 = u_4$ is the position that is equal to the green-coloured stage in the tree.

Note first that the sets $\{w_1, w_2\}$, $\{w_3, w_4\}$, $\{w_1, w_4\}$ and $\{w_3, w_2\}$ form cuts because every root-to-sink path has to pass through w_1 and w_3 on the upper half of the CEG and through w_2 and w_4 on the bottom half. Now let for example $W = \{w_1, w_2\}$ and let X_W be a cut-variable indicating whether a unit has passed through w_1 or through w_2. Then using Theorem 2, we obtain that both $Y_{w_3} \perp\!\!\!\perp Y_{w_0} \mid X_W$ and $Y_{w_4} \perp\!\!\!\perp Y_{w_0} \mid X_W$. Translated into our underlying random variables, this simply implies that both $Z \perp\!\!\!\perp Y \mid X = 0$ and $Z \perp\!\!\!\perp Y \mid X = 1 \vee X = 2$. In fact, because W is also a fine cut, we know that $Z \perp\!\!\!\perp Y \mid X$.

In a BN model, the above assumptions can be represented by a collider DAG $Y \to X \leftarrow Z$. However, this DAG does not fully code all hypotheses expressed in (4.9) but simply states that Y and Z are independent. So we would need a context-specific BN rather than a BN to capture the extra information that Z does not simply depend on X but that instead Z groups the levels of the random variable X into two categories $X = 0$ and $X \neq 0$ such that both the events $X = 1$ and $X = 2$ have the same implications on the system.

We can see above that assumptions which in a context-specific BN need to be communicated using formulae and *non-graphical* information can trans-

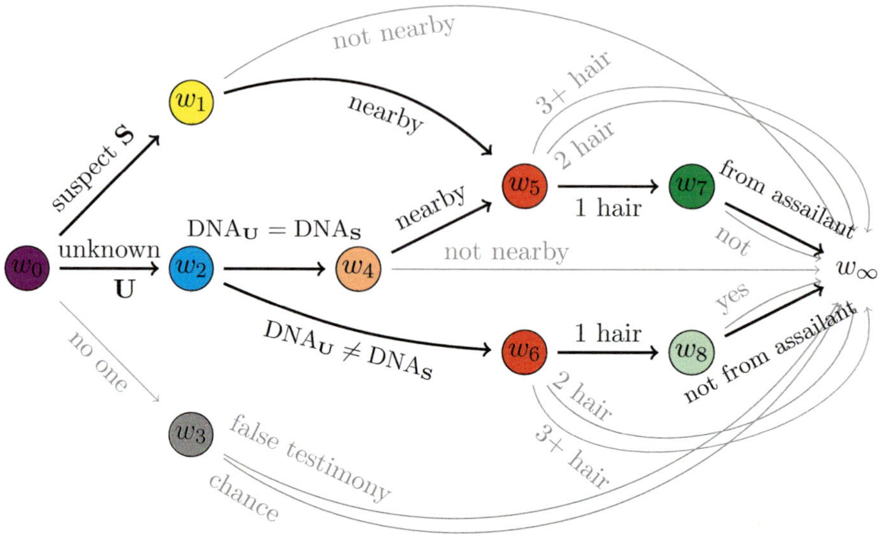

FIGURE 4.4
A CEG with corresponding staged tree given in Figure 1.1 and discussed in Section 1.2. See also Example 4.10.

parently be communicated *directly* when using a Chain Event Graph rather than a DAG.

These extra assumptions often have very straightforward interpretations in applications, as in the examples below.

Example 4.10 (A legal case)**.** *The CEG in Figure 4.4 is an alternative representation of the staged tree Figure 1.1 from Section 1.2 representing a rape case. The root-to-sink paths corresponding to the defence and the prosecution hypotheses are again depicted black in this illustration while complementary unfoldings have been greyed out.*

Here, the vertices $W' = \{w_1, w_2, w_3\}$ form a position cut. By Theorem 2, vertex-random variables which are downstream of this cut will not depend on the value of the root-variable Y_{w_0} given that we have observed W'. As a consequence, for instance whether we found one or more hair on the victim is independent of whether the suspect, an unknown person or no one is guilty of having assaulted the victim. Similarly, whether an assault happened or not is independent of who was guilty given that a hair was found. Further, the quantity of hair found is independent of the suspect's and an unknown man's DNA being equal given that they were both nearby.

When we focus entirely on the subCEG given by the prosecution and defence cases then in this graph both $\{w_5, w_6\}$ and $\{w_7, w_8\}$ are cuts. In particular, the probability of a hair having been donated by the assailant is inde-

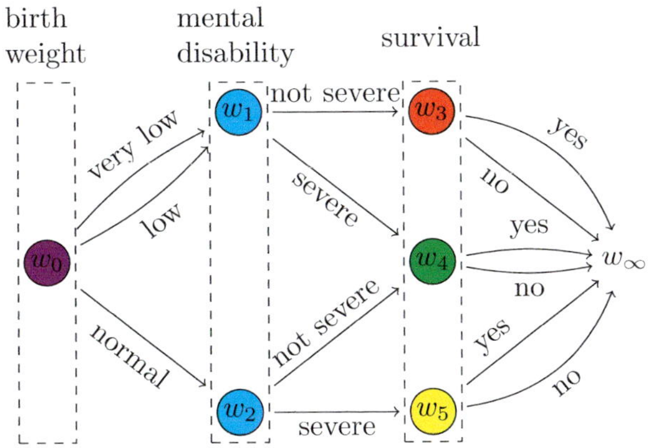

FIGURE 4.5
A CEG with corresponding staged tree given in Figure 1.2 from [6]. See Example 4.11 and Section 1.3.

pendent of any past unfoldings and is thus in particular independent of who the assailant actually was.

We see here that the CEG representation of the rape case can transparently help to communicate and validate the assumptions we have made when setting up this case in Section 1.2.

Example 4.11 (Cerebral palsy). *The CEG in Figure 4.5 was inferred to be the best fitting CEG of a dataset on cerebral palsy patients [6]. Its corresponding staged tree and the story behind this example were presented in Section 1.3: see also Figure 1.2.*

We can now again define vertex-random variables and, using Theorem 2, read from this CEG the following conditional independence information. First, we can see immediately that mental disability and birth weight are independent random variables. Second, we can further infer from the graph that if we know that the birth weight is either very low or low then the chance of survival of a patient depends only on the severity of his or her mental disability. The same holds if the birth weight is normal. Using the cuts of this CEG we can thus group patients into two birth-weight categories, (very) low *and* normal, *rather than three. For patients within these two groups their respective future unfoldings are the same.*

The class of conditional independence relationships we can investigate by examining the results of conditioning on an arbitrary intrinsic event in a CEG is orders of magnitude larger than that possible using an alternative BN model, even when these representations are equivalent. For instance, in the cerebral-palsy model from Example 4.11 an intrinsic event to condition on could be whether a patient has a low or very low birth weight. In the CEG in Figure 4.5 this simply corresponds to conditioning on the vertex-centred event $\Lambda(w_1)$ which we have interpreted as 'not normal birth weight'. But such conditioning—X_{w_0} taking one variable *or* another—cannot be represented through an alternative BN description of the problem. So within the usual BN architecture we do not even consider this type of conditioning. In Chapter 6 we will show how such model simplifications can be found in practice, using model search algorithms.

Because the CEG representation of a model is so rich, it is usually very hard to provide an exhaustive list of *all* of its associated conditional independence statements. For simple CEGs, Theorem 2 can be reformulated such that the conditional independence constraints read from a CEG are necessary and sufficient for the underlying probability mass function: so the assumptions read from (fine) cuts are the only constraints on a model [106]. An alternative way of directly specifying the families of distributions in these models is via generating functions: we present this approach in the section below.

4.2 CEG statistical models

In this section we will provide a formal analysis of statistical models represented by CEGs following [43, 47]. In particular, we will split the question Q2 we asked at the beginning of this chapter into the following three subquestions:

Q2.1 Given a probability distribution, can this be represented by a CEG?

Q2.2 Given a CEG, can we represent the underlying model by a different CEG instead?

Q2.3 Given two CEGs, can we tell whether these represent the same model and can we transform one into the other using graphical operations?

Figure 4.6 illustrates the relationship between the three queries above.

An answer to the first question requires a formal specification of a CEG in terms of a set of probability distributions with certain properties. These are depicted as hypersurfaces inside the probability simplex (or tetrahedron) in the figure. An answer to question Q2.1 would give us a method to check whether or not a given distribution is a point lying on one of these surfaces. Some basic concepts from algebraic geometry will provide this method in Section 4.2.1. The second question addresses the notion of statistical equivalence, also known as distribution equivalence or, in particular, Markov-equivalence in BNs: see Section 2.3 and for instance [50]. In the figure below, statistically equivalent staged trees are depicted within clouds of the same colour as the model they represent, in dark or light gray, respectively. An answer to question Q2.2 would enable us to find different members within these clouds. The third question Q2.3 asks for a number of graphical operations which traverse a class of statistically equivalent CEGs, so move around between staged trees in equally coloured clouds in the figure. We will answer these last two questions

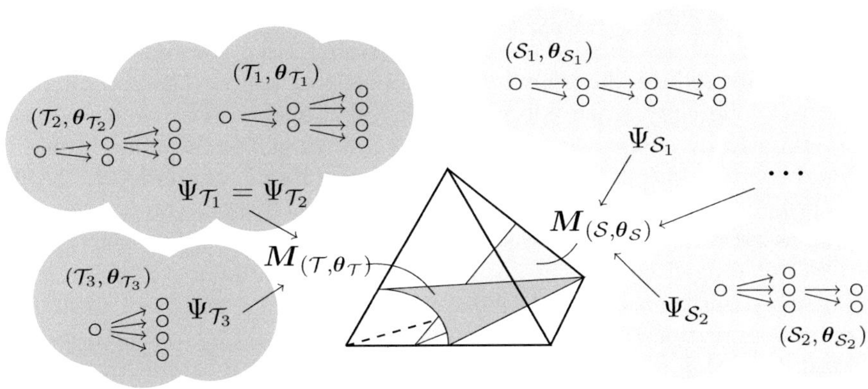

FIGURE 4.6
A summary of the contents of Section 4.2. Here, $\boldsymbol{M}_{(\mathcal{T}, \boldsymbol{\theta}_{\mathcal{T}})}$ and $\boldsymbol{M}_{(\mathcal{S}, \boldsymbol{\theta}_{\mathcal{S}})}$ are two different staged tree models inside a probability simplex, represented by graphs $(\mathcal{T}, \boldsymbol{\theta}_{\mathcal{T}})_i$ and $(\mathcal{S}, \boldsymbol{\theta}_{\mathcal{S}})_j$, respectively, $i = 1, 2, 3$; $j = 1, 2$. These graphs sometimes share the same parametrisation $\Psi_{\mathcal{T}_1} = \Psi_{\mathcal{T}_2}$, and all graphs whose parametrisations map to the same surface represent the same model.

in Section 4.2.2 where the *swap* operator will enable us to move around one single cloud in the picture—representing what we call a 'polynomial' equivalence class—and in Section 4.2.3 where the *resize* operator will enable us to change between clouds within a statistical equivalence class.

It is most straightforward to present the answers to the above queries in terms of staged trees in the development below. Because by Theorem 1 a staged tree and its CEG represent the same model, we can then directly translate results from one framework to the other. All results presented here have been adapted from [47] with extra examples presented in [43].

Recall from Section 3.1 that by (3.2), a staged tree model is a discrete and parametric statistical model as in Definition 2.1 from Chapter 2. The parametrisation of such a model is induced by the multiplication rule along the root-to-leaf paths of a corresponding staged tree (or CEG) representation. Whenever two such parametrisations have the same image, they specify the same model. We will thus work with the following definition.

Definition 4.12 (Statistical equivalence). *We will call two staged tree representations* $(\mathcal{T}, \boldsymbol{\theta}_{\mathcal{T}})$ *and* $(\mathcal{S}, \boldsymbol{\theta}_{\mathcal{S}})$ *of the same model* statistically equivalent, *so if and only if* $\boldsymbol{M}_{(\mathcal{T}, \boldsymbol{\theta}_{\mathcal{T}})} = \boldsymbol{M}_{(\mathcal{S}, \boldsymbol{\theta}_{\mathcal{S}})}$.

In answering the three queries Q2.1–Q2.3 above, we will provide a complete characterisation of the full statistical equivalence class of a given model.

The study of these statistical equivalence classes is an important one. The first reason for this is computational: CEGs constitute a massive model space to explore, as we will see in the subsequent chapters. By identifying a single representative within an equivalence class of model representations and a priori selecting across these representatives rather than the full class, we can dramatically reduce the search effort across this space. The second reason concerns coherence: when adopting a Bayesian approach in model selection, two statistically equivalent models (those always giving the same likelihood) should be given the same prior distribution over its parameters [50]. To apply this principle, it is essential to know when two CEGs make the same distributional assertions. The third reason is inferential: just like a Bayesian network, a CEG or staged tree has a natural causal extension. In particular, causal discovery algorithms can be applied to CEGs to elicit a putative causal ordering between various associated variables. Clearly a necessary condition for a causal deduction to be made is that this deduction is invariant to the choice of one representative within a statistical equivalence class. So again we need to be able to identify equivalence classes of a hypothesized causal CEG in order to perform these algorithms. Details of this procedure will be given in Section 8.4.

4.2.1 Parametrised subsets of the probability simplex

By construction, every staged tree or CEG model can be specified as the image of a monomial map $\Psi_{\mathcal{T}} : \boldsymbol{\theta}_{\mathcal{T}} \to \boldsymbol{M}_{(\mathcal{T},\boldsymbol{\theta}_{\mathcal{T}})}$ which can be read from a graphical representation of the model. The distributions in this image are then given as vectors of atomic probabilities, parametrised using the edge labels of the graph. The components of these vectors are strictly positive and sum to unity. Staged tree models are thus very special subsets of a probability simplex as depicted in Figure 4.6.

We illustrate this geometric interpretation with the revisiting of an example familiar from Section 3.1.

Example 4.13 (A model on three atoms). *Consider a toss of a possibly biased coin which is repeated only if the outcome of the first toss is* heads. *This is the setting from Example 3.2 where we now toss the same coin twice. The staged tree $(\mathcal{T},\boldsymbol{\theta}_{\mathcal{T}})$ in Figure 4.7 represents this setting.*

We label the edge e_1 in this tree by the probability of heads, denoted $\theta = \theta(e_1)$, and—because the transition probabilities from the same vertex sum to unity—the edge e_2 by the probability of tails, so $\theta(e_2) = 1-\theta$. Because the two tosses are independent, the transition probabilities from v_1 are the same as from v_0, or rather $\theta(e_3) = \theta(e_1)$ and $\theta(e_4) = \theta(e_2)$. Thus v_0 and v_1 are in the same stage and $(\mathcal{T},\boldsymbol{\theta}_{\mathcal{T}})$ is a staged tree which is not square-free.

The atomic probabilities of the possible single outcomes of this experiment are then $p_{\boldsymbol{\theta},\mathcal{T}}(\lambda_1) = \theta^2$, $p_{\boldsymbol{\theta},\mathcal{T}}(\lambda_2) = \theta(1-\theta)$ and $p_{\boldsymbol{\theta},\mathcal{T}}(\lambda_3) = 1-\theta$, for $\boldsymbol{\theta} = (\theta, 1-\theta)$. The coin-toss model is therefore the set of distributions $p_{\boldsymbol{\theta},\mathcal{T}}$ which assign the above probabilities to three atoms, for all positive probabilities θ of heads,

$$\boldsymbol{M}_{(\mathcal{T},\boldsymbol{\theta}_{\mathcal{T}})} = \left\{\boldsymbol{p}_{\theta,\mathcal{T}} = \left(\theta^2,\, \theta(1-\theta),\, 1-\theta\right) \mid \theta \in (0,1)\right\} \tag{4.10}$$

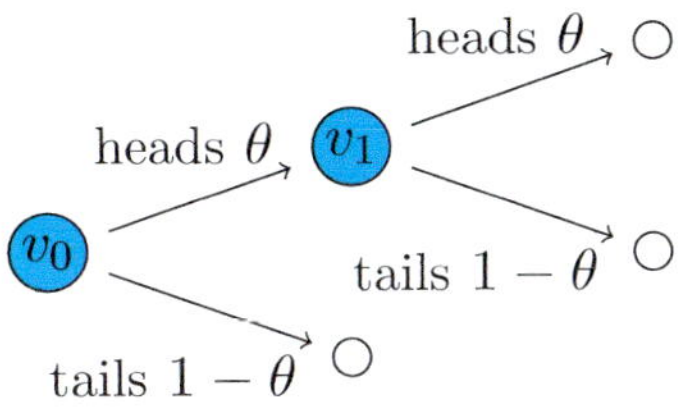

FIGURE 4.7
A non square-free staged tree $(\mathcal{T}, \boldsymbol{\theta}_\mathcal{T})$ which represents a coin-toss model. For an explanation see Example 4.13.

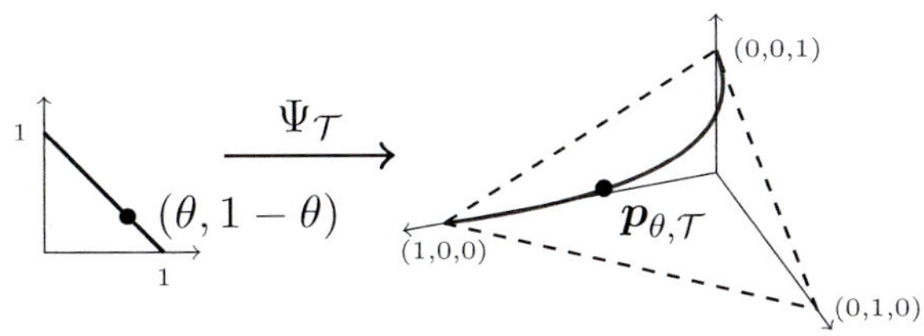

FIGURE 4.8
The parametrisation $\Psi_\mathcal{T} : \theta \mapsto \boldsymbol{p}_{\theta,\mathcal{T}}$ belonging to $(\mathcal{T}, \boldsymbol{\theta}_\mathcal{T})$ from Figure 4.7 maps a parameter to an element in the model $\boldsymbol{M}_{(\mathcal{T},\boldsymbol{\theta}_\mathcal{T})}$ from Example 4.13.

as in Definition 2.1.

All components of a vector $\boldsymbol{p}_{\boldsymbol{\theta},\mathcal{T}} \in \boldsymbol{M}_{(\mathcal{T},\boldsymbol{\theta}_\mathcal{T})}$ *are strictly positive and sum to unity, thus every vector does indeed correspond to a distribution over the three atoms. In particular, the staged tree model for this coin-toss experiment is a parametrised set of points within a space of discrete probability distributions. Its parametrisation is the bijective map* $\Psi_\mathcal{T} : \theta \mapsto \boldsymbol{p}_{\theta,\mathcal{T}}$.

From a geometric point of view, the model $\boldsymbol{M}_{(\mathcal{T},\boldsymbol{\theta}_\mathcal{T})}$ *is a parametric curve in three-dimensional space. We draw this curve in Figure 4.8. Now a distribution factors according to this staged tree model if and only if it is a point on the curve. For instance, in the illustration above a distribution*

$$\big(P_{\boldsymbol{\theta}}(\text{heads} \wedge \text{heads}), P_{\boldsymbol{\theta}}(\text{heads} \wedge \text{tails}), P_{\boldsymbol{\theta}}(\text{tails})\big) = (0.49, 0.21, 0.3) \quad (4.11)$$

can describe our experiment because it is a point on the curve, namely the image of $(\theta, 1-\theta) = (0.7, 0.3)$ *which assigns probability* 0.7 *to the outcome* heads *and probability* 0.3 *to the outcome* tails. *In contrast, the point* $(0.24, 0.06, 0.7)$ *does not lie on this curve: it is a distribution factorising according to the saturated model in Example 3.2 but not according to the staged tree model (4.10).*

The ideas in the example above can be extended to characterise any type of CEG or staged tree model as a parametrised subset of a probability simplex—or alternatively as the solution set of a system of polynomial equations and inequalities. These polynomial equations can be given in terms of *odds ratios* and thus allow for a very straightforward interpretation within the statistical model [43]. In the language of algebraic geometry, staged tree models are *semi-algebraic sets.* In particular, they are varieties specified by odds-ratio equations of degree 2 intersected with the positivity and sum-to-1 constraints of a probability simplex.

This provides an answer to Q2.1: whenever a distribution fulfills a collection of odds-ratio equations and positivity constraints then it factorises according to a staged tree. This result enables us to avoid having to find an inverse map to a given parametrisation in order to check whether a specific point lies on a model. Instead we can simply plug in the values of the given distribution into a set of equations and see whether these are fulfilled.

We will not give any technical proofs to these results here but we instead illustrate this finding in a number of examples below.

Example 4.14 (The saturated model). *Let $(\mathcal{T}, \boldsymbol{\theta}_\mathcal{T})$ be a probability tree which is trivially staged—with every stage containing precisely one situation—and which has $n \in \mathbb{N}$ root-to-leaf paths. Then the corresponding probability mass function, and hence the corresponding vector of atomic probabilities $\boldsymbol{p}_{\boldsymbol{\theta},\mathcal{T}}$ can take any value within the $n-1$-dimensional probability simplex*

$$\Delta^\circ_{n-1} = \{\boldsymbol{p} \in \mathbb{R}^n \mid \sum_{i=1}^{n} p_i = 1 \text{ and } p_i \in (0,1) \text{ for all } i = 1, \ldots, n\} \tag{4.12}$$

as in Section 2.2.2. So this staged tree model fills the entire probability simplex.

We will henceforth call $\boldsymbol{M}_{(\mathcal{T},\boldsymbol{\theta}_\mathcal{T})} = \Delta^\circ_{n-1}$ a saturated *tree model, and we call the trivially staged probability tree a* saturated tree *as in Section 3.4.1.*

Saturated models are the most simple forms of staged tree models. More interesting structure arises when we consider for instance staged trees with independence structure.

Example 4.15 (The independence model). *Consider again the staged tree and CEG from Examples 4.2, 4.4 and 4.8. All of these graphs represent the binary independence model*

$$\begin{aligned} \boldsymbol{M}_{X \perp\!\!\!\perp Y} &= \{p(x,y) = p_1(x)p_2(y) \mid p_1, p_2 \text{ probability distributions}; x, y = 0, 1\} \\ &= \big\{\big(\theta_0\tau_0, \theta_0\tau_1, \theta_1\tau_0, \theta_1\tau_1\big) \mid (\theta_0, \theta_1), (\tau_0, \tau_1) \in \Delta^\circ_{2-1}\big\}. \end{aligned} \tag{4.13}$$

The parametrisation we use here can be read directly from the graph in Figure 4.2: see also (4.7). We draw the resulting model as a hypersurface inside the three-dimensional probability simplex Δ°_{4-1} in Figure 4.9a.

An alternative non-parametric characterisation of this independence model can be given in terms of the odds-ratio equation

$$\frac{p_1}{p_3} = \frac{p_2}{p_4} \tag{4.14}$$

whose solution set $\{\boldsymbol{p} = (p_1, p_2, p_3, p_4) \in \mathbb{R}^4 \mid \boldsymbol{p} \text{ fulfils } (4.14)\} \cap \Delta^\circ_{4-1}$ inside the probability simplex equals precisely the model (4.13) above.

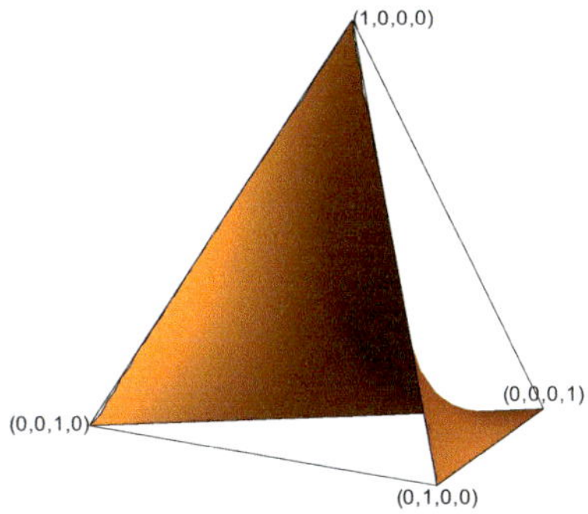

(a) The independence model $\boldsymbol{M}_{X \perp\!\!\!\perp Y}$ from Example 4.15.

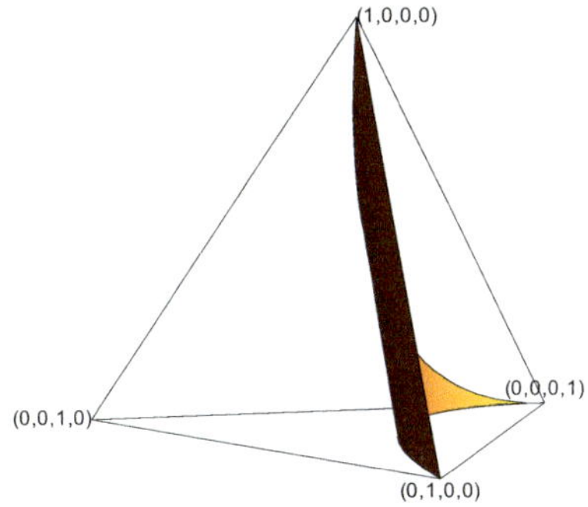

(b) The staged tree model $\boldsymbol{M}_1$ from (4.15a) and (4.16a).

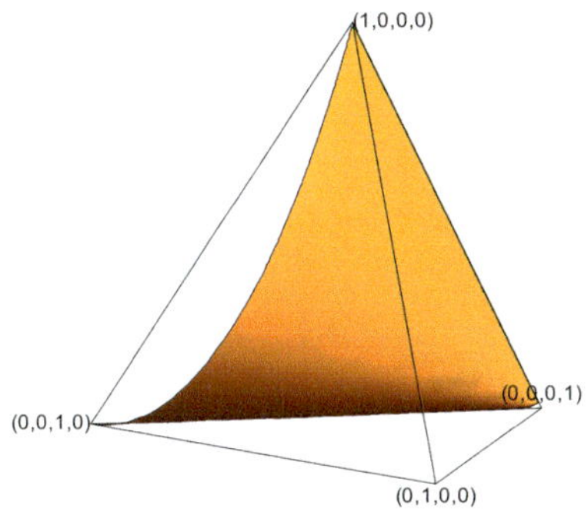

(c) The staged tree model $\boldsymbol{M}_2$ from (4.15b) and (4.16b).

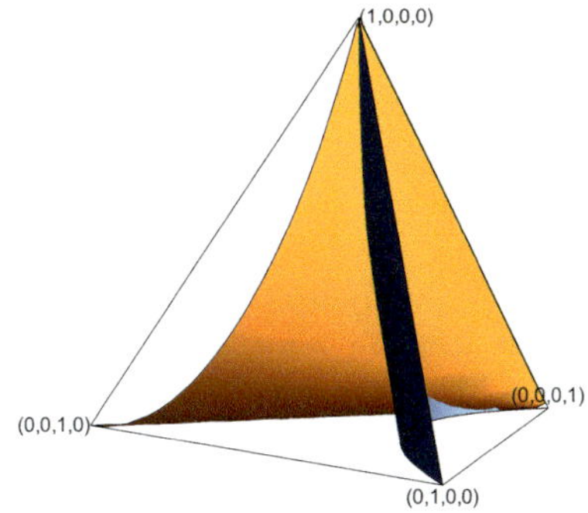

(d) The staged tree model $\boldsymbol{M}_3$ from (4.15c) and (4.16c).

FIGURE 4.9
Staged tree models on four atoms, analysed in Examples 4.15 and 4.16.

The idea of characterising a discrete statistical model in terms of the solution set of polynomial equations and inequalities has originated from the research area of *Algebraic Statistics* [29, 41, 75, 79] and has already been hugely successful in the characterisation of graphical models, including Bayesian networks [36, 38, 39].

We are currently exploring the algebro-geometric properties of staged tree models. In particular, a complete characterisation of all staged tree models on four atoms is now available [43]. We repeat a few special cases below.

Example 4.16 (Three staged tree models on four atoms)**.** *Consider the three staged trees in Figure 4.10. These are graphical representations of the respect-*

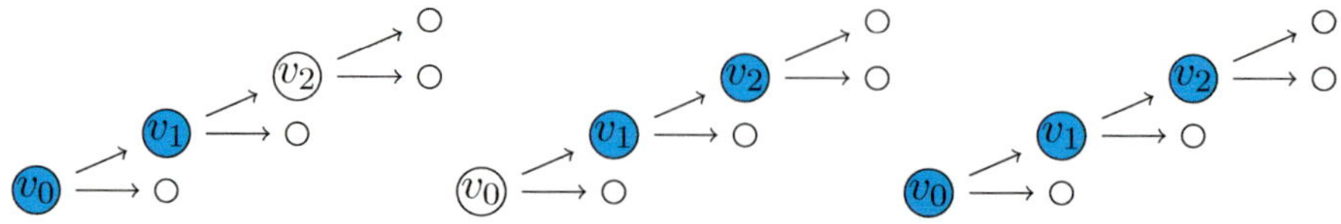

(a) The staged tree $(\mathcal{T}, \boldsymbol{\theta}_\mathcal{T})_1$. (b) The staged tree $(\mathcal{T}, \boldsymbol{\theta}_\mathcal{T})_2$. (c) The staged tree $(\mathcal{T}, \boldsymbol{\theta}_\mathcal{T})_3$.

FIGURE 4.10
Three staged trees on four atoms which represent models analysed in Example 4.16 and depicted in Figures 4.9b to 4.9d, respectively.

ive models

$$\boldsymbol{M}_1 = \left\{\left(\theta_0^2\tau_0, \theta_0^2\tau_1, \theta_0\theta_1, \theta_1\right) \mid (\theta_0, \theta_1), (\tau_0, \tau_1) \in \Delta_{2-1}^{\circ}\right\} \tag{4.15a}$$
$$\boldsymbol{M}_2 = \left\{\left(\theta_0\tau_0^2, \theta_0\tau_0\tau_1, \theta_0\tau_1, \tau_1\right) \mid (\theta_0, \theta_1), (\tau_0, \tau_1) \in \Delta_{2-1}^{\circ}\right\} \tag{4.15b}$$
$$\boldsymbol{M}_3 = \left\{\left(\theta_0^3, \theta_0^2\theta_1, \theta_0\theta_1, \theta_1\right) \mid (\theta_0, \theta_1) \in \Delta_{2-1}^{\circ}\right\} \tag{4.15c}$$

where we use the shorthand $\boldsymbol{M}_i = \boldsymbol{M}_{(\mathcal{T}, \boldsymbol{\theta}_\mathcal{T})_i}$ for the model represented by the staged tree $(\mathcal{T}, \boldsymbol{\theta}_\mathcal{T})_i$, for $i = 1, 2, 3$.

An alternative specification of these parametric statistical models can again be given as the solution sets of the following systems of equations

$$(p_1 + p_2)p_4 = p_3(p_1 + p_2 + p_3) \tag{4.16a}$$
$$p_1p_3 = p_2(p_1 + p_2) \tag{4.16b}$$
$$(p_1 + p_2)p_4 = p_3(p_1 + p_2 + p_3) \text{ and } p_1p_3 = p_2(p_1 + p_2) \tag{4.16c}$$

for $(p_1, p_2, p_3, p_4) \in (0, 1)^4$ such that $p_1 + p_2 + p_3 + p_4 = 1$. From this new characterisation is it immediately obvious that the third model $\boldsymbol{M}_3 = \boldsymbol{M}_1 \cap \boldsymbol{M}_2$ is equal to the intersection of the first two models. This is because every distribution that fulfils both (4.16a) and (4.16b) also fulfils (4.16c), and vice versa. We depict this result in Figures 4.9b to 4.9d. Note also that the staged tree $(\mathcal{T}, \boldsymbol{\theta}_\mathcal{T})_3$ representing that model contains the stage structure of both $(\mathcal{T}, \boldsymbol{\theta}_\mathcal{T})_1$ and $(\mathcal{T}, \boldsymbol{\theta}_\mathcal{T})_2$. So there is a direct graphical analogue to this geometric result.

We can now move on to specifying all staged trees representing the same model as above, so those giving rise to the same solution set of odds-ratio equations and representing the same subsets of a probability simplex.

4.2.2 The swap operator

Unlike the situation we outlined for BNs in Section 2.3 where model representations which make equivalent distributional assumptions can be elegantly characterized through DAGs sharing the same pattern, sadly no such simple common graphical representation is available for staged trees or CEGs. However, we can instead define a polynomial in the edge labels of a staged tree which provides a natural algebraic index for a class of equivalent staged trees, to be used as an analogue of the essential graph in this more general framework.

Because CEGs are parametric models and their parametrisation is given in terms of the unknown edge labels of a tree graph, it is highly advantageous to address this issue and to define the polynomial below in a *symbolic* framework. Here, we do not assign numerical values to these parameters but treat them as indeterminates which—only at the very end of our analysis—can be interpreted as primitive probabilities.

The idea of associating a polynomial to a graph, or more generally to a parametric statistical model, in order to answer inferential queries has originated from three very intuitive insights.

First, inference in statistical models often relies on calculations based on evaluating polynomials and rational functions. In particular, joint probabilities can be calculated by multiplying conditional probabilities, conditional probabilities can be calculated using fractions and marginal probabilities are calculated by summing over joint probabilities. So a thorough analysis of the polynomials involved in these types of calculations promises to give us a better understanding of the mechanisms underlying a statistical model. This approach has already been successful in BN modelling [25].

Second, generating functions of discrete statistical models and their generalisations are often of polynomial form [80]. Because these functions can identify certain classes of models, this suggests that a special type of polynomial associated to a probability tree model might be used to generate distributions which factorise according to the graphs representing this model: this is indeed the case as we will show below.

Third, computer scientists have been using DAGs whose inner nodes correspond to summation or multiplication operations for representing and evaluating values of polynomials for a long time. These *arithmetic circuits* or *sum-product nets* are fast inferential tools [26]. So there is a straightforward and natural link between polynomials and graphs, our graphical models.

Definition 4.17 (Interpolating polynomial)**.** *Let* $(\mathcal{T}, \boldsymbol{\theta}_\mathcal{T})$ *be a staged tree. We define the* interpolating polynomial *associated to the staged tree as the formal sum of all atomic probabilities,*

$$c_\mathcal{T}(\boldsymbol{\theta}) = \sum_{\lambda \in \Lambda(\mathcal{T})} \prod_{e \in E(\lambda)} \theta(e). \tag{4.17}$$

This is a special case of the more general definition provided in [47].

By definition, the interpolating polynomial of a staged tree is simply a sum over all products of edge labels along its root-to-leaf paths—without assigning any numerical values to these.

The event tree then yields a straightforward way to parenthesise an interpolating polynomial as follows. For every floret $\mathcal{F}_v$ where $v \in V$ is the parent of a leaf, we sum all components of its parameter vector $\boldsymbol{\theta}_v$ and multiply the result by its parent label $\theta(\mathrm{pa}(v), v)$. We then sum the result over the parent's labels $\boldsymbol{\theta}_{\mathrm{pa}(v)}$. We repeat this until all floret parameter vectors are recursively summed and until we arrive at the root $\mathrm{pa}(v) = v_0$. The interpolating polynomial can then be written in terms of a nested bracketing

$$c_{\mathcal{T}}(\boldsymbol{\theta}) = \sum_{v_1 \in \mathrm{ch}(v_0)} \theta(v_0, v_1) \Big(\sum_{v_2 \in \mathrm{ch}(v_1)} \theta(v_1, v_2) \; \ldots \; \Big(\sum_{v_k \in \mathrm{ch}(v_{k-1})} \theta(v_{k-1}, v_k) \Big) \Big) \tag{4.18}$$

where the index $k \in \mathbb{N}$ implicitly depends on the length of the root-to-leaf path $((v_0, v_1), (v_1, v_2), \ldots, (v_{k-1}, v_k))$. Every inner bracket in the sum above now includes a sum over the children of a vertex whose parent's children are summed in an outer bracket. In particular, the innermost brackets correspond to a sum over primitive probabilities assigned to leaves and the outermost to primitive probabilities belonging to the root. This bracketing thus inductively follows the nesting of the path-sigma algebra depicted in the event tree: compare also Section 4.1.1.

An important aspect of the result above is that it is reversible [47]. Not only can we easily read a polynomial from a probability tree but we can also construct an event tree from such a nested factorisation of a polynomial. In particular, a set of distributions can be represented by a staged tree or CEG if and only if there exists a parametrisation of the model such that the formal sum of their atomic probabilities can be written in terms of a nested bracketing as in (4.18) where certain subsums are identified with each other. This answers the question Q2.1 we posed at the beginning of this section and is the constructive counterpart to the result given in Section 4.2.1. So we can specify distributions which factor according to staged trees or CEGs either in terms of a characterisation of a given parametrisation (and hence the tree itself) or in terms of a set of equations and inequalities.

An algorithm for deciding whether such a bracketed expression of a given polynomial exists is now available [44].

The reversibility result above is important because whenever two staged trees have the same interpolating polynomial, they also share a common parametrisation and are thus statistically equivalent. We call this subclass of staged trees the class of *polynomially equivalent* staged trees.

Example 4.18 (Polynomial equivalence). *Consider the staged tree* $(\mathcal{T}, \boldsymbol{\theta}_{\mathcal{T}})$ *depicted on the left hand side in Figure 4.11. This tree has the associated interpolating polynomial*

$$\begin{aligned} c_{\mathcal{T}}(\boldsymbol{\theta}) &= \theta_0 + \theta_1\phi_2 + \theta_1\phi_2 + \theta_2\phi_1\sigma_1 + \theta_2\phi_1\sigma_2 + \theta_2\phi_1\sigma_3 + \theta_2\phi_2 \\ &= \theta_0 + \theta_1(\phi_1 + \phi_2) + \theta_2(\phi_1(\sigma_1 + \sigma_2 + \sigma_3) + \phi_2) \end{aligned} \tag{4.19}$$

where $\boldsymbol{\theta} = (\theta_0, \theta_1, \theta_2, \phi_1, \phi_2, \sigma_1, \sigma_2, \sigma_3)$ *is the vector of all parameters. This polynomial can be bracketed precisely in the nested form given in (4.18). Here,* θ_0*,* θ_1 *and* θ_2 *are associated to the root and are the outermost factors in*

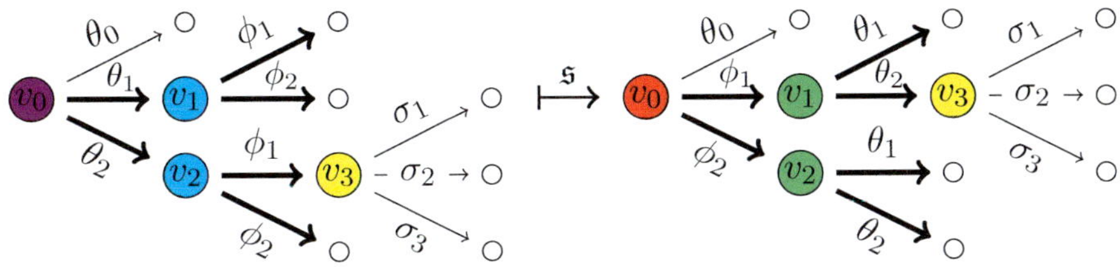

FIGURE 4.11
A swap $\mathfrak{s}$ transforms a staged tree $(\mathcal{T}, \boldsymbol{\theta}_{\mathcal{T}})$ into a polynomially equivalent staged tree $(\mathcal{S}, \boldsymbol{\theta}_{\mathcal{S}})$. See Examples 4.18 and 4.19.

that bracketing. ϕ_1 and ϕ_2 are factors in the inner brackets and label edges which are downstream from the root, and σ_1, σ_2 and σ_3 form a subsum in the innermost brackets because they label a leaf-floret.

It can be easily seen that given such a nested bracketing, we can simply reverse the algorithm described above and draw a staged tree from the interpolating polynomial: for every label we draw an edge, and these edges belong to the same floret whenever they are in the same subsum in the bracketing. The vertices at the endpoints of edges are then connected whenever they belong to the same term in the interpolating polynomial.

All polynomially equivalent staged trees arise from a reordering of a given nested bracketing as above. Each of these nestings corresponds to a different graphical representation within the same statistical equivalence class. So in order to traverse the class of polynomially equivalent staged trees using a simple local operator, we will need to answer the question: When can we reorder (4.18) without destroying the bracketing? And what is the graphical analogon to that arithmetic operation?

Interestingly, a local reordering of a nested polynomial representation as above is possible whenever we find two vertices in the corresponding graph which are in the same stage and have the same parent. We call these special subtrees *twins*—their root has identical children with the same parent.

For instance, both thick depicted subtrees in Figure 4.11 are twins, and the staged tree in Figure 4.2 is a twin. A twin in a CEG is always of the form given in Figure 4.12.

The interpolating polynomial of a twin factors into two terms

$$c_{\mathcal{T}'}(\boldsymbol{\theta}) = \Big(\sum_{e \in E(v_0)} \theta(e)\Big)\Big(\sum_{e' \in E(v)} \theta(e')\Big) \tag{4.20}$$

where v_0 is the root of the twin and v is any vertex in the stage $u \subseteq \mathrm{ch}(v_0)$ included in the twin. By the distributive law, the factorisation above can always be split such that we either sum over the $\theta(e)$-labels first or over the $\theta(e')$-labels first (or possibly also mixtures of these indeterminates). So in

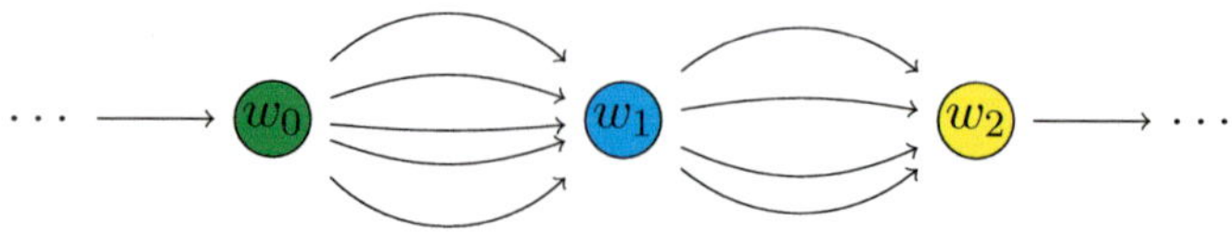

FIGURE 4.12
A twin inside a bigger CEG. Here, the swap operator reverses the order of events from $\Lambda(w_0) \prec \Lambda(w_1) \prec \Lambda(w_2)$ to $\Lambda(w_1) \prec \Lambda(w_0) \prec \Lambda(w_2)$ where $\prec$ denotes the partial order 'is downstream of'.

graphical terms, probability trees with this interpolating polynomial depict either outcomes associated with v_0 and its labels first, or outcomes associated with u and its labels. Both are valid representations of the same model. We call the operation which reorders the polynomial above and changes the order of parent and children in the twin a *swap*.

A swap is hence a local graphical operation on a staged tree which allows us to replace a certain two-level subtree of the original tree by one which has the same interpolating polynomial, and is thus polynomially equivalent, while leaving the rest of the original tree invariant.

Twins in staged trees correspond to local subgraphs representing (conditional) independence: compare Example 4.8. Our very plausible discovery is that for these independent events the order is reversible within a polynomial—and hence statistical—equivalence class, using the swap operator: see also Figure 4.12. This result is central to a causal analysis of these representations, as outlined in the final chapter of this book.

Example 4.19 (The swap operator)**.** *Consider again the setting in Example 4.18. The staged tree $(\mathcal{T}, \boldsymbol{\theta}_\mathcal{T})$ exhibits a twin around its blue-coloured stage $u = \{v_1, v_2\}$. This twin has been thick depicted in Figure 4.11. It has an associated interpolating polynomial that factors as $(\theta_1 + \theta_2)(\phi_1 + \phi_2)$ as in (4.20). As a consequence, there exist two different graphical representations of the underlying model: one which depicts the floret labelled by θ_1 and θ_2 first, and a different representation which depicts a floret labelled by ϕ_1 and ϕ_2 first. In fact, we can see this by reordering the bracketing of the interpolating polynomial $c_\mathcal{T}$ from (4.19) to:*

$$c_\mathcal{S}(\boldsymbol{\theta}) = \theta_0 + \phi_1(\theta_1 + \theta_2(\sigma_1 + \sigma_2 + \sigma_3)) + \phi_2(\theta_1 + \theta_2). \qquad (4.21)$$

Of course the two polynomials (4.19) and (4.21) have the same distributed form, so $c_\mathcal{T} = c_\mathcal{S}$. However, the two bracketings are different and they thus belong to different tree representations. The staged tree $(\mathcal{S}, \boldsymbol{\theta}_\mathcal{S})$ whose bracketing is given in this example is polynomially equivalent to $(\mathcal{T}, \boldsymbol{\theta}_\mathcal{T})$ whose polynomial representation was given above. We depict the new, polynomially equivalent representation of this model on the right hand side of Figure 4.11.

The operation which reorders the bracketing above and maps one tree onto the other is a swap $\mathfrak{s} : (\mathcal{T}, \boldsymbol{\theta}_\mathcal{T}) \mapsto (\mathcal{S}, \boldsymbol{\theta}_\mathcal{S})$.

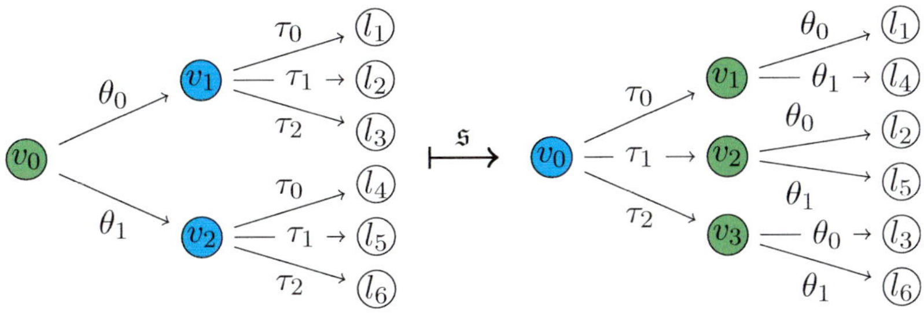

FIGURE 4.13
Two polynomially equivalent staged trees which are twins and can be transformed into each other using a swap. Note that whilst the labels of these trees are the same and the atomic probabilities are the same, the graphs are very different.

When applying a swap the description of a model in terms of an event tree can undergo far more radical changes than the example above suggests: numerous examples of this behaviour are given in [43]. Consider also Figure 4.13 which illustrates two polynomially equivalent staged trees that have very different graphs.

When leaving the symbolic framework and assigning numerical values to the edge labels of a staged tree, floret sum-to-1 conditions can be very different across two different but polynomially equivalent representations of the same model.

Example 4.20 (Sum-to-1 conditions)**.** *When interpreting the edge labels in Figure 4.11 as primitive probabilities, we note that the same parameters cannot necessarily take the same values across both representations. This is because even though the atomic probabilities across the two model representations are the same, they can arise from very different floret structures.*

For instance, in the staged tree $(\mathcal{T}, \boldsymbol{\theta}_{\mathcal{T}})$ *we require that both* $\theta_0+\theta_1+\theta_2 = 1$ *and* $\phi_1 + \phi_2 = 1$*. But in the polynomially equivalent tree* $(\mathcal{S}, \boldsymbol{\theta}_{\mathcal{S}})$ *we require that both* $\theta_0 + \phi_1 + \phi_2 = 1$ *and* $\theta_1 + \theta_2 = 1$*. Clearly, if these parameters should be* probabilities *then not all four equations can be true at the same time. However, we can think of these parameters as* potentials*—or* primitive probabilities*—so unnormalised marginal or conditional probabilities which can be assigned different values across different probability mass functions: compare [63]. In this case, both representations are staged trees with unambiguous interpretations.*

For instance, here we might have that $(\theta_0, \theta_1, \theta_2) = (0.2, 0.7, 0.1)$ *and* $(\phi_1, \phi_2) = (0.4, 0.6)$ *in* $(\mathcal{T}, \boldsymbol{\theta}_{\mathcal{T}})$ *and that* $(\theta_0, \phi_1, \phi_2) = (0.2, 0.7, 0.1)$ *and* $(\theta_1, \theta_2) = (0.4, 0.6)$ *in* $(\mathcal{S}, \boldsymbol{\theta}_{\mathcal{S}})$*. Naturally, both choices yield the same atomic probabilities in both representations.*

The subclass of all polynomially equivalent staged trees representing the same model—depicted as one cloud in our illustration in Figure 4.6—can be completely traversed using the swap operator [47]. This result gives a partial answer to questions Q2.2 and Q2.3 above.

The swap operator is a close tree analogue of an *arc reversal* in BNs [89]. These reversals, just like swaps, allow us to traverse the class of all graphical representations of the same model while renormalising (but not marginalising) the associated probability mass function. In particular, we can define an interpolating polynomial on a decomposable BN model using a clique-parametrisation, and then analyse its possible nested factorisations. As a result we find that the class of polynomially equivalent $\boldsymbol{X}$-compatible staged trees contains one statistically equivalent staged tree representation for every member of the class of all junction tree representations of that BN model [43]. So the polynomial equivalence class of a staged tree can be thought of as the 'junction tree equivalence class' of a Bayesian network.

4.2.3 The resize operator

The class of polynomially equivalent staged trees (or CEGs) is a proper subclass of the statistical equivalence class, as illustrated below. This is because the swap operator is defined on one given polynomial, arising from a fixed parametrisation of a model. Naturally, there may be very different parametrisations—and very different event trees—representing the same model. This motivates the introduction of a second operator on staged trees which will enable us to traverse the whole statistical equivalence class, and thus to fully answer the question Q2.3 we posed at the beginning of this section.

Example 4.21 (Statistical but not polynomial equivalence)**.** *Denote by* $\mathcal{F} = (v_0, \{e_1, \ldots, e_n\})$ *a floret with* $n \in \mathbb{N}$ *edges, and let* $\boldsymbol{\theta}_{\mathcal{F}} = \big(\theta(e_i) \mid i = 1, \ldots, n\big)$ *be an associated vector of edge labels. Then the* star *tree graph which is the labelled floret* $(\mathcal{F}, \boldsymbol{\theta}_{\mathcal{F}})$ *is a probability tree. Every model represented by such a star is not subject to any constraints, so is saturated:* $\boldsymbol{M}_{(\mathcal{F},\boldsymbol{\theta}_{\mathcal{F}})} = \Delta^{\circ}_{n-1}$.

But also the general saturated tree $(\mathcal{T}, \boldsymbol{\theta}_{\mathcal{T}})$ *from Example 4.14 was a valid representation of this model. As a consequence,* $(\mathcal{T}, \boldsymbol{\theta}_{\mathcal{T}})$ *and* $(\mathcal{F}, \boldsymbol{\theta}_{\mathcal{F}})$ *are very different probability trees but are still statistically equivalent. The labels of the saturated tree can be interpreted as conditional probabilities, while the labels of the star are products of these, or joint atomic probabilities. This implies that they have different associated parametrisations, and cannot be polynomially equivalent.*

An operation which contracts saturated subtrees of a staged tree and which contracts subgraphs rooted at vertices which are in the same position into single florets will not violate any specified atomic probabilities. This operation will thus always result in a staged tree representation of the same statistical model. We henceforth call such an operation and its inverse a *resize.*

Resizes are operations which enable us to change a given parametrisation

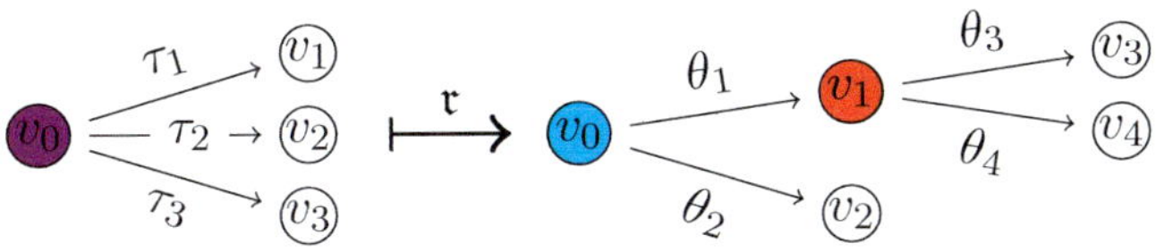

FIGURE 4.14
A resize operator on staged trees as in Example 4.22.

of a model and so move from one cloud to the other in the illustration provided in Figure 4.6.

Example 4.22 (The resize operator)**.** *The floret on the left hand side of Figure 4.14 has edges labelled by* (τ_1, τ_2, τ_3) *and can be transformed into a binary tree, depicted on the right hand side of that figure. This new tree has edge labels* (θ_1, θ_2) *and* (θ_3, θ_4)*.*

The resize $\mathfrak{r}$ *transforming one graph into the other applies a monomial map* $(\tau_1, \tau_2, \tau_3) \mapsto (\theta_1\theta_3, \theta_1\theta_4, \theta_2)$ *on these edge labels whose inverse is rational*

$$(\theta_1, \theta_2, \theta_3, \theta_4) \mapsto (\tau_1 + \tau_2, \tau_3, \frac{\tau_1}{\tau_1 + \tau_2}, \frac{\tau_2}{\tau_1 + \tau_2})$$

and can be calculated based on the rules for joint and conditional probabilities.

*It can be easily checked that sum-to-*1 *conditions are preserved under this operation. In fact, if* $\tau_1 + \tau_2 + \tau_3 = 1$ *then also* $\theta_1 + \theta_2 = \tau_1 + \tau_2 + \tau_3 = 1$ *and* $\theta_3 + \theta_4 = {}^{\tau_1}/_{\tau_1+\tau_2} + {}^{\tau_2}/_{\tau_1+\tau_2} = 1$*. Vice versa, if* $\theta_1 + \theta_2 = \theta_3 + \theta_4 = 1$ *then also* $\tau_1 + \tau_2 + \tau_3 = \theta_1(\theta_3 + \theta_4) + \theta_2 = 1$*. So again different local sum-to-*1 *conditions do not affect the atomic probabilities.*

The resize is particularly strong in terms of tightening a graphical representation of a model or splitting florets into more detailed depictions of an unfolding of events. In particular, we find that in every statistical equivalence of a staged tree there is a *minimal* representation given by a staged tree with no saturated subtrees bigger than florets—corresponding to a 'minimally sufficient parametrisation' of the model—as well as a *maximal* representation where every floret is binary. These special representations are not unique [43].

The resize operation is again the direct tree analogue to a well-known operation in Bayesian networks. When representing a BN model by a DAG then there are no conditional independence constraints between variables within the same clique: see Section 2.3. Hence, in probabilistic inference on the class of decomposable BNs the random variables belonging to a clique are often merged into one joint variable and the DAG can then be replaced by a junction tree. In terms of the associated probability mass function, a product of conditional probabilities is hereby replaced by a joint probability. This is precisely the operation we perform on saturated subtrees of a staged tree when resizing. So both in DAGs and in staged trees, uninformative subgraphs can be replaced by more compact graphical structures. In particular, a resize performed on a saturated subtree of an $\boldsymbol{X}$-compatible staged tree is directly equivalent to imposing a clique-parametrisation as above.

4.2.4 The class of all statistically equivalent staged trees

The swap and the resize operator are sufficiently powerful to enable us to traverse the whole equivalence class of a given staged tree, using both a symbolic description of the model in terms of the interpolating polynomial and at the same time performing intuitive transformations on the underlying graph.

Theorem 3 (Statistical equivalence). *Two square-free staged trees* $(\mathcal{T}, \boldsymbol{\theta}_{\mathcal{T}})$ *and* $(\mathcal{S}, \boldsymbol{\theta}_{\mathcal{S}})$ *are statistically equivalent if and only if there exists a map* $\mathfrak{m} : (\mathcal{T}, \boldsymbol{\theta}_{\mathcal{T}}) \mapsto (\mathcal{S}, \boldsymbol{\theta}_{\mathcal{S}})$ *which is a finite composition of swaps and resizes.*

A proof of this rather technical result can be found in [47]. Further examples of the two operators are provided in [43, 44]. Swaps enable us to move around within the clouds depicted in Figure 4.6 whilst resizes enable us to move around between equally coloured clouds in that figure. Whenever there are no swaps or resizes between two staged trees—so for instance between two differently coloured clouds in that figure—then these trees represent different models, and their parametrisations map to different subsets of the probability simplex.

Taking a composition of the swap and resize operator enables us to overcome the limitations of using one or the other transformation exclusively. Indeed, using swaps we can change the order of certain events and discover which renormalisations of an underlying probability mass function are possible in a polynomial equivalence class without reparametrising the model. Using resizes, we can reparametrise and tighten uninformative saturated subtrees. Resizing these subtrees might then create new twins and might enable us to swap subtrees which were formerly spuriously fixed in the polynomial equivalence class. In terms of more general parametric models, the algebraic operations underlying Theorem 3 simply ascertain that two graphs representing the same statistical model are reparametrisations of each other. The value of the result above lies in the corresponding powerful graphical operations which are very intuitive.

Consider an illustration below which is motivated by the study of a real-world problem.

An example of statistical equivalence: Prison radicalisation

We give a very simple and naïve version of an ongoing study about the nature of the process of radicalisation within prisons [17]. Our focus in the analysis presented here is to identify those prison inmates who are most likely to engage in specific criminal organizations.

In order to illustrate this point using the theory developed above, we have restricted our analyses to consider only four variables describing the problem, three of which are explanatory variables. The following have been chosen because they are often hypothesised as playing a key role in the process of radicalisation:

Gender X_g: a binary variable distinguishing between *male* and *female*,

Age X_a: a binary variable differentiating between *young* inmates (age < 30) and *mature* inmates (age ≥ 30),

Offence X_o: a nominal variable with three categories, namely *violence* against person, robbery, burglary, theft; *drug offence*; and *other*,

Risk X_r: a binary variable distinguishing between individuals at *high* or *low* risk of radicalisation.

In this type of study actual risk assessments are generally very coarse. This is why we will base the model we develop here on expert judgements [24, 48, 60, 73, 94] and cluster inmates into three different radicalisation classes of risk, assuming that mature male inmates and young female inmates have the same respective probabilities of committing each type of crime. We also assume that the highest risk prisoners are the young male inmates who have committed violence-related crimes. The lowest risk prisoners are female and mature male inmates who committed non-drug or non-violence type of crimes, and young female inmates who were convicted of drug offence. As a result, there are only three non-trivial stages in a compatible staged tree representation of this setting which respects the variable ordering $X_\text{g} \prec X_\text{a} \prec X_\text{o} \prec X_\text{r}$. We depict these stages red, blue and green in Figure 4.15a.

Importantly, when setting up a BN model for the problem above then the only DAG representation of this setting is given by a complete graph on the four variables X_g, X_a, X_o and X_r. This graph does not convey any of the rich context-specific information we are given above and is thus a very bad choice of representation. However, a CEG model or staged tree representation enables us to construct a compact and very expressive graph to represent all of these context-specific statements.

Using the expert judgement above, we can thus draw out one possible representation of the problem. We will now analyse the statistical equivalence class of this staged tree using Theorem 3 in order to draw out alternative staged trees representing the same model. Observe first that the subtree emanating from the root with situations v_0, v_1 and v_2 and their emanating edges is saturated. This part of the graph does thus not include any stage information, or constraints which specify the model. Using a resize operator as in Section 4.2.3, it is thus sensible to tighten the graphical representation given here and to transform this subtree into a single floret. This floret has now four emanating edges, labelled *young male*, *mature male*, *young female* and *mature female*. In this process we have thus merged the two random variables *age* and *gender* which were given a priori to describe the system. After resizing, we now explain the radicalisation process using a new random variable *demographic group* X_{d} which is associated to the root vertex and has the four states given above. This alternative staged tree representing the same model is depicted in Figure 4.15b.

Two further resizes to tighten the description of our system would be possible on the saturated subtree around the situation v_7, or on the two subtrees emanating from v_4 and v_5 which are in the same position. We will refrain from applying these resizes in order to keep the stratification of the original staged tree representation intact. This has the advantage that we can now analyse putative reorderings of the random variables $X_{\mathrm{d}} \prec X_{\mathrm{o}} \prec X_{\mathrm{r}}$ using the swap operator.

Consider first the vertices v_4 and v_5 on the second level of the original staged tree: these form a twin in the resized tree because they are in the same stage and now also have the same parent, namely the root in the resized tree. Using a swap operation on this red-coloured stage, we can see that the underlying staged tree model allows for representations depicting demographics before type of offence $X_{\mathrm{d}} \prec X_{\mathrm{o}}$ or, alternatively, reversing this order $X_{\mathrm{o}} \prec X_{\mathrm{d}}$ for the group of mature males and young females. So for this demographic subgroup, the two random variables X_{d} and X_{o} are independent and we cannot infer from the model that age and gender explain the type of offence these individuals commit.

The blue- and green-coloured twins given by v_8 and v_9, v_{10} and v_{11}, v_{13} and v_{14}, and v_{17} and v_{18}, respectively, on the third level of the staged tree can be analysed in a very similar fashion: for young males who committed drug or other offences, the respective probabilities of radicalisation are the same as for mature males and young females who committed more serious violence or drug-related crimes. For mature females who committed drug or other offences, the radicalisation probabilities are the same as for young females or mature males who committed other crimes. For none of these groups can we infer an unambiguous order on offence and radicalisation from the model. As a consequence, in none of these cases can the offence be interpreted as 'causing' a radicalisation: see Chapter 8. In fact, the model suggests that the interplay of these problem variables is far more complex than depicted in the stratification.

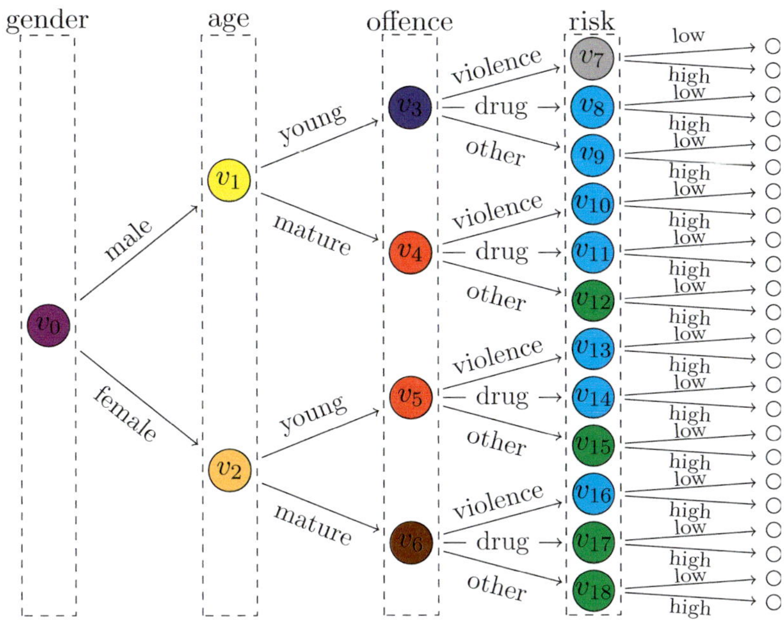

(a) A staged tree elicited from expert judgements.

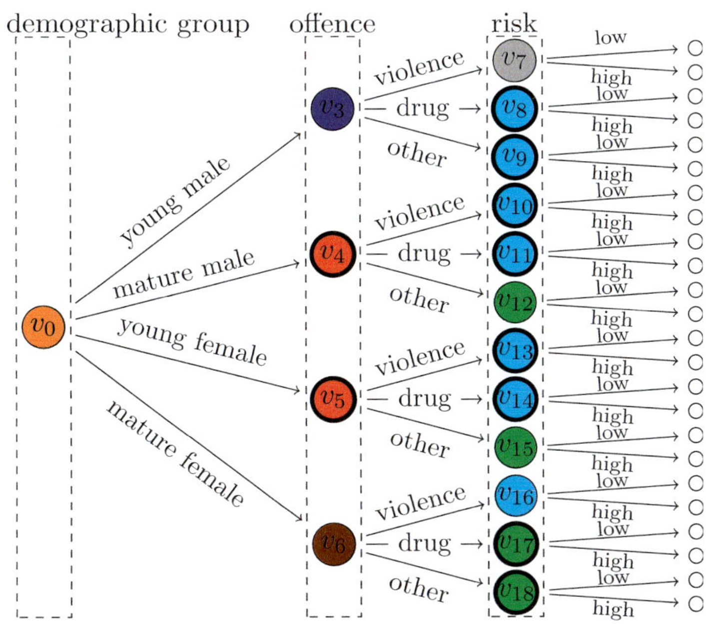

(b) A staged tree arising from a resize. Twin situations are encircled thick.

FIGURE 4.15
Two statistically equivalent staged trees for the prison radicalisation model.

4.3 Exercises

Exercise 4.1. *Consider the snakebite CEG in Figure 3.5b. Depict the subgraphs* $\mathcal{C}_{\Lambda(w_1)}$ *and* $\mathcal{C}_{\Lambda(w_3)}$ *separately. Determine the intersection of the edge and vertex sets of these subgraphs. Show that this graphical intersection is not equal to the graph of the event-intersection* $\mathcal{C}_{\Lambda(w_1)\cap\Lambda(w_1)}$. *Why?*

Exercise 4.2. *Find events in a CEG of your choice which are not intrinsic. Find events which are intrinsic. Describe all of these as unions and intersections of vertex- and edge-centred events.*

Exercise 4.3. *Show that the set of intrinsic events is closed under intersection. Show that the map* $\Phi : \mathcal{C}_\Lambda \mapsto \Lambda$ *is a homomorphism from the set of subgraphs of a CEG to the path-sigma algebra of that CEG, preserving the intersection operation. Determine a set of events for which* Φ *is an isomorphism.*

Exercise 4.4. *Using Theorem 2, find a list of (conditional) independence statements on the vertex-random variables depicted in the staged tree in Figure 4.15 representing the prison radicalisation model in Section 4.2.4.*

Exercise 4.5. *Using the swap operator, determine the polynomial equivalence classes of all* $\boldsymbol{X}$*-compatible staged trees encountered in this chapter.*

Exercise 4.6. *Draw all statistically equivalent staged trees discussed in the radicalisation process in Section 4.2.4, using both the swap and the resize operator.*

5

Estimation and propagation on a given CEG

In the previous chapter we have demonstrated how useful a CEG can be for embodying structural information. In particular, this graph is able to depict many types of hypothesised dependence relationships between different types of events. It also enables us to embellish these hypotheses by annotating its edges with probabilities. These can then be used to generate a full probability model over the atoms of this space detailing how and how likely different events might unfold. In practice, we are then often faced with two situations: first, we might need to estimate the probabilities on a given CEG on the basis of a real-world dataset; second, we might observe the (partial) history of one individual in the modelled population and then need to revise these probabilities in the light of that evidence.

We will show in this chapter how to tackle both of these problems. The first section below assumes that we are given a graph and need to find its associated edge probabilities, mirroring the Bayesian inference methods and conjugacy analyses presented in Chapter 2. We hereby distinguish between fully observed cases and those where data is missing, both by design or at random. The second section then builds on these methods and hands us powerful algorithms to tackle the case where we observe a subsystem of our model and want to update our beliefs about the unfoldings of that subpopulation, often a single individual. The methods we present here work for any type of CEG presented in Chapter 3.

5.1 Estimating a given CEG

A natural next question to ask is: *What happens when the embellishing edge probabilities of a CEG are uncertain, or how do we learn these from data?* This is an inferential question which encourages us to place a probability distribution on the different collections of edge probabilities—so on the vectors of floret labels for each individual stage. The most straightforward way to do this is to perform a conjugate analysis as in Section 2.1. We show below how this can be achieved. In particular, in this section our focus will be on the situation when the graph $\mathcal{C} = \mathcal{C}(\mathcal{T})$ of a CEG $(\mathcal{C}(\mathcal{T}), \boldsymbol{\theta}_{\mathcal{T}})$ is treated as known and so fixed but its associated edge probabilities $\boldsymbol{\theta}_{\mathcal{T}}$ are uncertain or need to be estimated. In the following chapter we will then extend these methodologies so that they apply to the case when the graph $\mathcal{C}$ of the CEG is itself uncertain.

5.1.1 A conjugate analysis

The development in this section is completely analogous to the one presented in Section 2.1.2, here tailored to CEG models. We follow the development of [33, 34] and set up some simple notation first.

Suppose we have elicited a Chain Event Graph $\mathcal{C}$ with a set of $K = \#U_{\mathcal{C}}$ stages. Throughout the next two chapters we will for simplicity of notation number these stages $u_1, \ldots, u_K$ and denote their associated floret parameter vectors by $\boldsymbol{\theta}_i = \boldsymbol{\theta}_{u_i}$ for $i = 1, 2, \ldots, K$. We further denote the edges of any vertex in a stage u_i by $E(u_i) = \{e_{i1}, \ldots, e_{iK_i}\}$ and their labels by $\theta_{ij} = \theta(e_{ij})$ for $j = 1, \ldots, K_i$ and $i = 1, \ldots, K$. We assume here that the floret probability vectors $\boldsymbol{\theta}_i$, for $i = 1, \ldots, K$, are mutually independent a priori. When the CEG is also a BN then this assumption corresponds to the almost ubiquitous assumption of local and global independence as discussed in Section 2.3. As in Chapter 3, we denote again by $\boldsymbol{\theta} = (\theta_{ij} \mid j = 1, \ldots, K_i; i = 1, \ldots, K)$ the vector of all edge labels of the CEG.

The root-to-sink paths of the CEG will now play the role of the categories $\omega_1, \ldots, \omega_n$ that observed units fall into, and these are determined by sequences of edges e_{ij} that units pass along. Furthermore, the probability distribution $p_{\boldsymbol{\theta},\mathcal{C}}$ we want to make inference on is now uniquely determined by the vector $\boldsymbol{\theta}$, so we aim to estimate the components of this vector. In the Bayesian framework, $\boldsymbol{\theta}$ is a random variable.

We henceforth assume that we observe a complete sample $\boldsymbol{Y} = \boldsymbol{y}$ where for every unit from the population modelled by the CEG at hand we know which category it falls into. To be precise, we henceforth denote this vector as $\boldsymbol{y} = (\boldsymbol{y}_0, \ldots, \boldsymbol{y}_K)$ where each $\boldsymbol{y}_i = (y_{i1}, \ldots, y_{iK_i})$ is a vector whose component y_{ij} denotes the number of units that arrive at stage u_i and then pass along the edge e_{ij} in the associated floret, for $j = 1, \ldots, K_i$ and $i = 1, \ldots, K$.

Provided that the sampling experiment was properly randomised, just as in Section 2.1.1, standard statistical theory assures us that given a floret parameter vector $\boldsymbol{\theta}_i$, each observation $\boldsymbol{Y}_i$ has a Multinomial distribution Multi$(\sum_{j=1}^{K_i} y_{ij}, \boldsymbol{\theta}_i)$ whose density we denote $f_i(\boldsymbol{y}_i|\boldsymbol{\theta}_i)$ for all $i = 1, \ldots, K$. Explicitly and in direct analogy to the development presented for general Bayesian prior-to-posterior analyses, the likelihood $f(\boldsymbol{y}|\boldsymbol{\theta})$ associated to the model represented by the CEG $\mathcal{C}$ with edge probabilities $\boldsymbol{\theta} = \boldsymbol{\theta}_{\mathcal{C}}$ will then take a separable form given by the product of the likelihoods of all probability vectors associated to the stages $u_1, \ldots, u_K$:

$$f(\boldsymbol{y}|\boldsymbol{\theta}) = \prod_{i=1}^{K} f_i(\boldsymbol{y}_i|\boldsymbol{\theta}_i) = \prod_{i=1}^{K} \prod_{j=1}^{K_i} \theta_{ij}^{y_{ij}}. \tag{5.1}$$

A conjugate analysis is now straightforward. For this assume, again as we also usually do for a BN model and as motivated in Section 2.1.3, that each of the floret parameter vectors $\boldsymbol{\theta}_i$ has a Dirichlet prior distribution $\text{Dir}(\boldsymbol{\alpha}_i)$ with parameter $\boldsymbol{\alpha}_i = (\alpha_{i1}, \ldots, \alpha_{iK_i})$. The prior distribution assigned to the whole vector $\boldsymbol{\theta}$ is thus chosen to be:

$$f(\boldsymbol{\theta}) = \prod_{i=1}^{K} \frac{\Gamma(\sum_{j=1}^{K_i} \alpha_{ij})}{\prod_{j=1}^{K_i} \Gamma(\alpha_{ij})} \prod_{j=1}^{K_i} \theta_{ij}^{\alpha_{ij}}. \tag{5.2}$$

We can now show that the Bayesian analysis for CEGs is closed under sampling. In particular, the Dirichlet prior distribution (5.2) can be updated given the Multinomial likelihood in (5.1) in order to obtain the posterior distribution of the parameter vector using Bayes' formula (2.1). Explicitly, for any given CEG $\mathcal{C}$ with parameter vector $\boldsymbol{\theta} = \boldsymbol{\theta}_{\mathcal{C}}$ and data vector $\boldsymbol{Y} = \boldsymbol{y}$ obtained from complete random sampling, we have the posterior

$$\begin{aligned} f(\boldsymbol{\theta}|\boldsymbol{y}) \propto f(\boldsymbol{y}|\boldsymbol{\theta}) f(\boldsymbol{\theta}) &= \prod_{i=1}^{K} \frac{\Gamma(\sum_{j=1}^{K_i} \alpha_{ij})}{\prod_{j=1}^{K_i} \Gamma(\alpha_{ij})} \prod_{j=1}^{K_i} \theta_{ij}^{y_{ij}+\alpha_{ij}-1} \\ = \prod_{i=1}^{K} f(\boldsymbol{\theta}_i|\boldsymbol{y}_i) &= \prod_{i=1}^{K} \frac{\Gamma(\sum_{j=1}^{K_i} \alpha_{ij+})}{\prod_{j=1}^{K_i} \Gamma(\alpha_{ij+})} \prod_{j=1}^{K_i} \theta_{ij}^{\alpha_{ij+}-1} \end{aligned} \tag{5.3}$$

where $\boldsymbol{\alpha}_{i+} = \boldsymbol{\alpha}_i + \boldsymbol{y}_i$ in the notation from Section 2.1. So in this conjugate analysis, the parameter vector $\boldsymbol{\theta}_i$ associated with each stage u_i can be learned independently and has a Dirichlet posterior distribution $\text{Dir}(\boldsymbol{\alpha}_{i+})$ for all $i = 1, \ldots, K$.

A big advantage of an analysis that is closed under sampling is that the marginal likelihood of the corresponding model can be written in closed form. In Section 6.2, we will use this property to quickly compare competing models. Explicitly, the marginal likelihood is

$$\begin{aligned} f(\boldsymbol{y}) = \int_{\Theta} f(\boldsymbol{y}|\boldsymbol{\theta}) f(\boldsymbol{\theta})\, \mathrm{d}\boldsymbol{\theta} &= \int_{\Theta} \prod_{i=1}^{K} \frac{\Gamma(\sum_{j=1}^{K_i} \alpha_{ij})}{\prod_{j=1}^{K_i} \Gamma(\alpha_{ij})} \prod_{j=1}^{K_i} \theta_{ij}^{y_{ij}+\alpha_{ij}-1}\, \mathrm{d}\boldsymbol{\theta} \\ &= \prod_{i=1}^{K} \frac{\Gamma(\sum_{j=1}^{K_i} \alpha_{ij})}{\Gamma(\sum_{j=1}^{K_i} \alpha_{ij+})} \prod_{j=1}^{K_i} \frac{\Gamma(\alpha_{+ij})}{\Gamma(\alpha_{ij})}. \end{aligned} \tag{5.4}$$

In particular, the logarithmic form of the marginal likelihood can therefore be expressed as a sum over log-Gamma functions of the hyperparameters associated with the prior and posterior distributions:

$$\log f(\boldsymbol{y}) = \sum_{i=1}^{K} \left(\log \Gamma(\overline{\alpha}_i) - \log \Gamma(\overline{\alpha}_{i+})\right) - \left(\sum_{j=1}^{K_i} \log \Gamma(\alpha_j) - \log \Gamma(\alpha_{j+})\right) \tag{5.5}$$

where we use the shorthand $\overline{\alpha}_i = \sum_{j=1}^{K_i} \alpha_{ij}$ for all $i = 1, \ldots, K$.

Although such a log-Gamma function is not a trivial mathematical function to be manually computed there are various software packages which can evaluate this function precisely and extremely quickly, for instance in statistical software such as R [82]. This enables us to explore the additive form of (5.5) for the design of efficient algorithms not only for model selection as presented in Section 6.2 but also for propagation as presented in Section 5.2. This has of course also been embedded in our own R-package ceg [18].

As a final note, observe that under complete random sampling the posterior parameters $\alpha_{ij+} = \alpha_{ij} + y_{ij}$ are simple linear functions of the prior and the data. This simple formula enables us to use the modelling hypotheses encoded within a CEG $\mathcal{C}$ to make strong inferences. In particular, we have shown in Section 2.1.2 that the inverse of the posterior variance—sometimes called the posterior *precision*—of each of the components of θ_{ij} increases linearly in $\overline{\boldsymbol{y}_i}$ for each posterior fixed mean vector $\boldsymbol{\mu}_{i+}$. If we have a sufficient count of units arriving at each of the stages—which is often automatic if the CEG has only a small number of stages—then we can estimate the probabilities on the florets of each stage accurately even with a moderately large dataset. This advantage of the CEG is in sharp contrast to the saturated model where we need samples at least ten times the number of atoms in the space before we can hope this to be so. This is one of the reasons why the embellishment of a probability tree through its stage structure is so valuable: we can use the associations hypothesised by the elicited stage structure of $\mathcal{C}$ to sharpen our inferences, decreasing the variance of our estimates.

Of course this sharpening is entirely dependent on the hypotheses embedded in $\mathcal{C}$ being valid. If the CEG is not in the right class of models, this precision will be unfounded and the associated estimates false. On the other hand because by assuming we have a graph $\mathcal{C}$ we can obtain good estimates of our probability distributions, we can usually see if the data contradicts these assumptions. This process can be done either directly using diagnostics or indirectly as part of a model selection algorithm. The latter is developed in the next chapter and illustrated in the one after.

5.1.2 How to specify a prior for a given CEG

Suppose we have elicited a CEG $\mathcal{C}$ and we have a dataset $\boldsymbol{Y} = \boldsymbol{y}$ of the root-to-sink paths taken by a complete random sample of individuals from the population we model using $\mathcal{C}$. *How can we set up an appropriate prior density over the stage probabilities associated with* $\mathcal{C}$*?* We are here looking for a motivation for using Dirichlet distributions $\text{Dir}(\boldsymbol{\alpha})$ in the previous section and for a recipe we can follow when setting their hyperparameters $\boldsymbol{\alpha}$, for instance in (5.2).

If possible then we should elicit each of the prior densities associated with the different stages directly from domain experts. There are many techniques available to do this: a review is given in [74]. However, even when the CEG of a given population can be treated as known this very precise elicitation exercise can be enormously time consuming. Especially when we have a rich data source available to us, it is often expedient instead to perform a conjugate analysis. Results such as those in [84] show that if the shape of the prior we choose is reasonably close to the one we might have obtained by such a careful elicitation and the sample data is sufficiently rich to concentrate the likelihood around a small neighbourhood of the parameters then very little is lost by approximating the inference by using a conjugate method rather than expert elicitation. We saw in the previous section that one family of conjugate distributions to use is a set of independent Dirichlet distributions, one for each of the probability vectors of the stages of a CEG. In Section 2.1 we outlined some very useful properties of the Dirichlet distribution. These ensure that if the atoms of a space are Dirichlet distributed then so are the margins and the conditionals associated with that space. In fact, product Dirichlets share similar—if slightly weaker—properties, provided we set the values of the hyperparameters appropriately. We show how to do this below.

It is helpful to first consider the special case when a CEG represents the saturated model on n atoms, where every stage consists of just one situation. We have seen in Section 4.2 that after application of a resize operator, this model is most simply represented by a star staged tree consisting only of a single root with n edges connecting it to n leaves. Suppose on the associated floret parameter vector we place a Dirichlet distribution $\text{Dir}(\boldsymbol{\alpha})$ with hyperparameter $\boldsymbol{\alpha} = (\alpha_1, \alpha_2, \ldots, \alpha_n)$. The results in Sections 2.1.1 and 2.1.3 tell us that $\boldsymbol{\alpha} = \overline{\alpha}\boldsymbol{\mu}$, so that the prior means $\boldsymbol{\mu}$ and the parameter $\overline{\alpha}$ fully specify this prior. Prior means can be elicited from a domain expert simply by asking what fraction of the population they expect to pass along each edge. The other parameter $\overline{\alpha} = \sum_{j=1}^{n} \alpha_j$ is called the *effective sample size*. This is a measure of the strength with which an expert holds their beliefs about these expected prior means. It can be elicited for instance by asking how large a population they base their beliefs about the prior means on. The larger $\overline{\alpha}$ is for fixed μ_j, the smaller are the expert's variances on each of the components of the associated probability vector, $j = 1, \ldots, n$. In particular, this parameter thus controls the weight we assign to the expert prior beliefs compared to the strength of evidence conveyed by the data.

For this saturated model the posterior $\text{Dir}(\boldsymbol{\alpha}_+)$ after observing numbers $\boldsymbol{y} = (y_1, y_2, \ldots, y_n)$ of units arriving at the n different leaves is simply given by $\alpha_{j+} = \alpha_j + y_j$ for all $j = 1, 2, \ldots, n$. One way of thinking of the role of the hyperparameter α_j is thus that it is a measure of the (possibly non-integer) number of *phantom* units that a priori the client believes will end up at the jth leaf, so how many will fall into that category of the population. It is this interpretation of α_j that will give us the rationale for setting hyperparameters across different CEGs: see Section 6.1. To make a fair comparison we will then demand that—when choosing between two models—that the strength of prior information *within* each competing model is the same in this sense.

We have seen in Section 4.2 that there are a number of statistically equivalent representations of this saturated model that are not given by a star. These can be obtained using the inverse resize operator. Ideally, in any given fixed scenario we would like the corresponding hyperparameter settings of our product Dirichlet prior to give the same posterior as the one we obtained above. Then it would not matter which of these data-indistinguishable representations we might use. In fact, when assuming all the Dirichlet floret probability vectors are a priori independent it is easy to check that this will imply an identical prior—and so posterior—on the Dirichlet prior on the atoms given above, for all these statistically equivalent models. As we have noted in Section 2.1, it is only the Dirichlet distribution that enjoys this property: see [33].

Now we can move on to the problem of setting a conjugate prior density for the general CEG $\mathcal{C}$. The most direct way to do this is to simply elicit each of the Dirichlet parameters associated with the floret parameter vector $\boldsymbol{\theta}_i$ on the ith stage, $i = 1, 2, \ldots K$, exactly as we would for the single Dirichlet above. Because we assume these parameter vectors to be independent, this elicitation task can be done for each stage individually. This would be our recommended way of eliciting the structure, especially when different experts contributed judgments to different parts of the tree—see e.g. [7, 99]—or when genuinely different strengths of information informed different parts of a tree.

However there is another way forward. Consider the special case where the client's prior information consists of having seen a certain number of units pass along various root-to-sink paths of the CEG. Then [51] argued for BNs and [33] argued for CEGs that the interpretation of the parameters above as counts of (possibly phantom) units would demand that the *mass conservation property* should hold for this CEG. This property simply demands that if we observe α_{ij} phantom units arriving at a vertex then $\sum_{i=1}^{K_i} \alpha_{ij}$ units should have arrived at its parent: so the population counts are preserved throughout the graph. For instance, if we set the prior effective sample size $\overline{\boldsymbol{\alpha}_0}$ on the floret associated with the the root then the number of these phantom units going along an edge is determined by the equation $\boldsymbol{\alpha}_0 = \overline{\alpha_0}\boldsymbol{\mu}_0$ where $\boldsymbol{\mu}_0$ again denotes the vector of elicited expected probabilities emanating from the root. It should therefore follow that the vector $\boldsymbol{\alpha}_0$ of the numbers of these phantoms proceeding along the different edges of the floret should be the vector of effective sample size parameters associated with the vectors of receiving situations. In particular, if a stage consisted of two such situations emanating from the root then the effective sample size parameter should simply be the sum of these two edge weights. We demonstrate this process of construction in the example presented in Section 5.1.3 below.

When prior information has the form discussed above then many useful properties follow. For instance, with complete random sampling where we see the passage of each sampled unit from the root to the sink of a CEG the mass conservation property is preserved a posteriori. Another small but useful consequence of using priors with mass conservation is that there is only one effective sample size parameter to set. If our prior information can be coded in this way then the effective sample size associated with any stage can be deduced from just one elicitation once the expected probabilities are fixed: often the root stage. If we are uncertain about this parameter we can investigate how inferences change as we vary this in a systematic way: see the full analysis presented in Chapter 7. Models with the mass conservation property are especially useful when addressing exploratory data analyses where we select a CEG on the basis of a complete dataset.

But what can we do when we are unable or unwilling to elicit the expected edge probabilities? Perhaps the most obvious way forward is to make all these expected edge probabilities emanating from any given stage equal: for instance by setting the phantom units arriving at each atom to one. Naturally, we cannot recommend this as a routine choice for many applied contexts. Eliciting even very rough and ready estimates of these probabilities from a client in our experience usually leads to better inferences. However, especially for demonstrating the general efficacy of different techniques and especially for model search, this is a convenient starting point for exploring the strength of relationships and the structure in a given dataset especially for SCEGs as introduced in Section 3.3. Of course when datasets are extremely large, even when this setting is not appropriate the posterior ensuing inferences will usually be robust to this choice due to properties referenced above. So then it could be an expedient choice. This setting used for any CEG will have set equal probabilities to all edges emanating from a particular position and also all probabilities of the edges emanating from a given situation. For the SCEG we also note that because of the symmetry of the underlying tree this assigns equal expected probabilities to all the atoms of the space—although this property may not hold when the CEG is not stratified. By setting a prior in this way we then simply need to elicit the CEG and a single effective sample size parameter as above.

5.1.3 Example: Learning liver and kidney disorders

To illustrate how to estimate a CEG from data and to further discuss how to set up a prior distribution consider now the toy example below which is an extended version of one presented in [108]. This will be our running example over the remainder of this chapter.

Patients diagnosed with liver and kidney disorders are often referred to a specialist by their general physician. We assume that each patient arriving at a specialised clinic can in theory be classified using three categories of liver and kidney dysfunction: minor, serious or critical. Regardless of this classification, every patient first receives clinical treatment (CT). If a patient with a minor disorder responds (Res) to the treatment, he will always make a full recovery (R). Otherwise, he will be designated for a semi-elective surgery (S_{se}). This can then result in lifetime monitoring (M), lifetime treatment (LT) or death

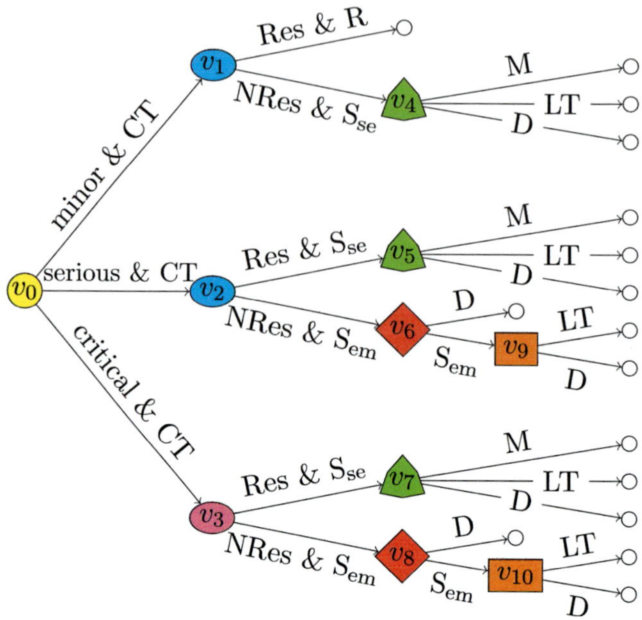

(a) A staged tree representation.

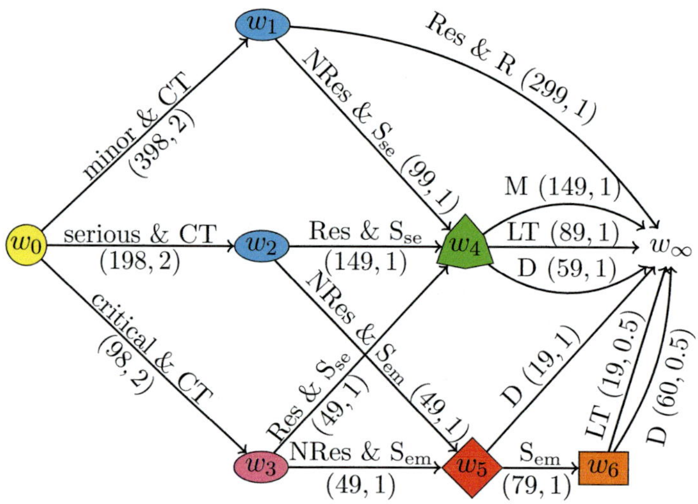

(b) A CEG representation.

FIGURE 5.1
The staged tree and CEG for the liver and kidney disorders model. The pair (y_{ij}, α_{ij}) associated with each outgoing edge e_{ij} from a position w_i in the CEG denotes the number of patients y_{ij} in the sample and the value of the hyperparameter α_{ij} in the phantom sample.

(D). A patient with a serious or critical disorder who responds to the clinical treatment will be also designated to a semi-elective surgery. However, if he does not respond (NRes) to the first treatment and he is still alive, he will then be admitted to an emergency surgery (S_{em}) whose consequences are either a lifetime of treatment or death. We depict these possible unfoldings of events in the staged tree $(\mathcal{T}, \boldsymbol{\theta}_{\mathcal{T}})$ in Figure 5.1a where every patient passes along one root-to-leaf path and ends up in one of the sixteen leaves. The edges of this tree are labelled by the shorthand we have introduced above.

We now make a number of assumptions. First that patients with minor or serious disorders have the same probability of responding to the clinical treatment. Second that the result of a semi-elective surgery has the same probability distribution for all patients. Third that patients with non-minor disorders have the same rate of survival along the clinical treatment and the emergency surgery. The positions in the staged tree are then

$$\begin{aligned} W_{\mathcal{T}} = \{&w_0 = \{v_0\}, w_1 = \{v_1\}, w_2 = \{v_2\}, w_3 = \{v_3\}, \\ &w_4 = \{v_4, v_5, v_7\}, w_5 = \{v_6, v_8\}, w_6 = \{v_9, v_{10}\}\}. \end{aligned} \tag{5.6}$$

These positions form the vertices of the corresponding CEG $\mathcal{C} = \mathcal{C}(\mathcal{T})$ depicted in Figure 5.1b. Here, all colours have been transferred from the tree to the CEG and the vertices w_1 and w_2 are in the same stage but not in the same position. The stage set of the CEG is thus

$$\begin{aligned} U_{\mathcal{C}} = \{&u_0 = \{w_0\}, u_1 = \{w_1, w_2\}, u_2 = \{w_3\}, \\ &u_3 = \{w_4\}, u_4 = \{w_5\}, u_5 = \{w_6\}\}. \end{aligned} \tag{5.7}$$

Now assume we have data available from a complete randomised trial following 694 patients that have experienced liver and kidney disorders. We will use these to estimate the vector $\boldsymbol{\theta}_{\mathcal{T}}$.

For illustrative purposes, we propose a prior where all stage probabilities have equal (uniform) edge hyperparameter and we set the effective sample size to six: so $\overline{\alpha} = 6$ and we let $\boldsymbol{\alpha}_0 = (2, 2, 2)$. Thus our prior belief—or the experience of the clinician—corresponds to having seen a phantom sample of six patients of which two each are assigned to minor, serious and critical clinical treatment. Using the conservation of mass property we discussed in the previous section, these parameters can be propagated through the CEG: we depict the resulting hyperparameters on each of the edges of the CEG in Figure 5.1b. For instance, since there are two edges arriving in the stage $u_1 = \{w_1, w_2\}$, the conservation of mass property yields that $\bar{\alpha}_1 = 4$ and $\boldsymbol{\alpha}_1 = (2, 2)$. As stage u_2 is constituted by a single position, it follows directly that $\bar{\alpha}_2 = 2$ and $\boldsymbol{\alpha}_2 = (1, 1)$. The other hyperparameters can be fixed in a similar way.

Using these parameters and the formulae we set up in Section 5.1.1, Table 5.1 now shows the prior and posterior probabilities associated with each stage of the CEG and Table 5.2 shows the respective prior and posterior 95% credible intervals and their prior and posterior variances.

TABLE 5.1
Prior and posterior probability distributions and data associated with each stage of the CEG depicted in Figure 5.1b.

Stage	Prior	$\boldsymbol{y}_i$, $i = 0, \ldots, 5$	Posterior	Posterior Mean
u_0	$\text{Dir}(2, 2, 2)$	$(398, 198, 98)$	$\text{Dir}(400, 200, 100)$	$(0.57, 0.29, 0.14)$
u_1	$\text{Dir}(2, 2)$	$(448, 148)$	$\text{Dir}(450, 150)$	$(0.75, 0.25)$
u_2	$\text{Dir}(1, 1)$	$(49, 49)$	$\text{Dir}(50, 50)$	$(0.5, 0.5)$
u_3	$\text{Dir}(1, 1, 1)$	$(149, 89, 59)$	$\text{Dir}(150, 90, 60)$	$(0.5, 0.3, 0.2)$
u_4	$\text{Dir}(1, 1)$	$(19, 79)$	$\text{Dir}(20, 80)$	$(0.2, 0.8)$
u_5	$\text{Dir}(0.5, 0.5)$	$(19, 60)$	$\text{Dir}(19.5, 60.5)$	$(0.24, 0.76)$

TABLE 5.2
Mean, 95% credible interval and variance of the prior and posterior probability distributions of stages u_0, u_1 and u_5 as in Table 5.1.

Distribution	Mean (95% credible interval)	Variance (10^{-4})
u_0 – Prior	$0.33(0.05, 0.72)$ $0.33(0.05, 0.72)$ $0.33(0.05, 0.72)$	$(317, 317, 317)$
u_0 – Posterior	$0.57(0.53, 0.61)$ $0.29(0.25, 0.32)$ $0.14(0.12, 0.17)$	$(3.4, 2.9, 1.7)$
u_1 – Prior	$0.5(0.09, 0.91)$ $0.5(0.09, 0.91)$	$(500, 500)$
u_1 – Posterior	$0.75(0.71, 0.78)$ $0.25(0.22, 0.29)$	$(3.1, 3.1)$
u_5 – Prior	$0.5(0.002, 0.998)$ $0.5(0.002, 0.998)$	$(1250, 1250)$
u_5 – Posterior	$0.24(0.16, 0.34)$ $0.76(0.66, 0.84)$	$(23, 23)$

Figures 5.2 and 5.3 depict the prior and posterior probability distributions of the stages u_0, u_1 and u_5. We can see here that the prior distributions have greater variance since their probability mass is more evenly distributed over the parameter space. Thus, for instance, the prior variances of each component of the stages u_0, u_1 and u_5 are, respectively, 0.03, 0.05 and 0.125 and their corresponding 95% credible intervals are, respectively, $(0.05, 0.72)$, $(0.09, 0.91)$ and $(0.002, 0.998)$. As we move from the root to the sink position in the CEG, the stage prior distributions become more dispersed: the variance is equal to 0.0317 for stage u_0 and to 0.125 for stage u_5.

In this example, just like discussed in the previous section, stages which are closer to the root have been visited more frequently and so tend to have a greater concentration of mass probability a posteriori. This is indeed sufficient to counterbalance any possible prior bias towards uniformity that a prior can impose over these stages: compare Figures 5.2b and 5.3a. In both cases the posterior variance of each component of these stages is less than $1,0 \times 10^{-3}$ and their 95% credible posterior intervals are narrow (of width 0.07).

In contrast, the stage u_5 which corresponds to the position w_6 is the most

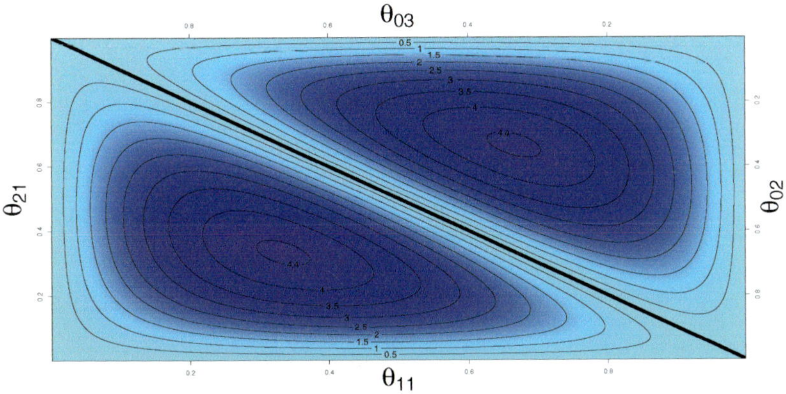

(a) Prior: $\boldsymbol{\theta}_0 \sim \text{Dir}(2, 2, 2)$.

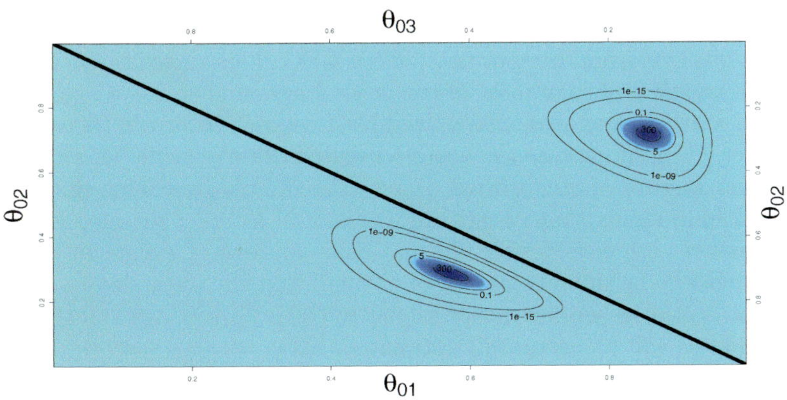

(b) Posterior: $\boldsymbol{\theta}_0 \sim \text{Dir}(400, 200, 100)$.

FIGURE 5.2
Prior and posterior probability distributions of the stage u_0 in the CEG depicted in Figure 5.1b. The upper right half of each picture corresponds to the joint probability distribution of the components associated with the diagnoses of serious (θ_{02}) and critical (θ_{03}) dysfunctions. The bottom left half of each picture corresponds to the joint probability distribution of the components associated with the diagnoses of minor (θ_{01}) and serious (θ_{02}) dysfunctions. Darker colour represents an area of higher joint probability.

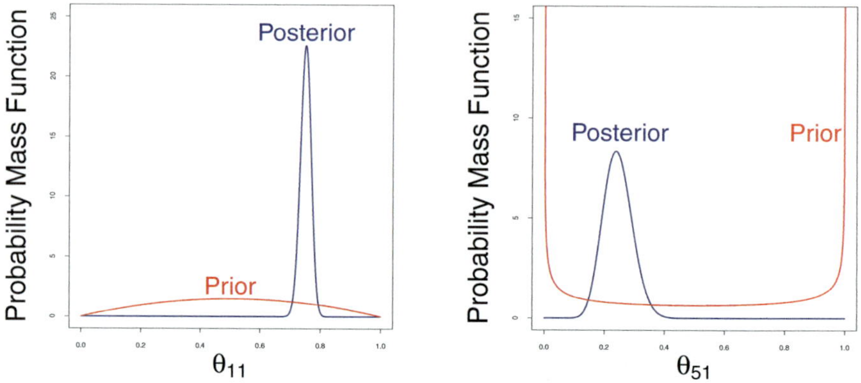

(a) Stage u_1: $\theta_{11} \sim \text{Beta}(0.5, 0.5)$ **(b)** Stage u_5: $\theta_{51} \sim \text{Beta}(19.5, 60.5)$

FIGURE 5.3
Prior and posterior probability distributions of stages u_1 and u_5 of the CEG depicted in Figure 5.1b.

distant position from the root node in our CEG. This stage is associated with a smaller sample size than most of the other stages and so has a higher variance (at $2,3 \times 10^{-3}$) and wider 95% credible posterior intervals (of width 0.18). However, even in this case the posterior distribution provides us with a reliable and clear picture of our process despite the sample size being only 79 out of the 694 individuals that constitute our overall sample: see also Figure 5.3b. So unless we are considering unusual combinations of events in populations and provided the number of stages is small and the data size moderate, the stage probabilities will be rather robust to the prior settings and the posterior probabilities will be accurately estimated.

5.1.4 When sampling is not random

We will now address some practical issues a user might face when modeling a process using a CEG and using data sources to inform this. Our first point is that we often do not have available to us a complete random sample. We have discovered in the analysis of a number of recent practical applications of CEG models that the likelihood of the data often still separates in a surprisingly wide number of cases. We list some of the more important of these classes below.

Conjugate ancestral sampling

Consider the case that a unit is observed until it arrives at a particular situation within a CEG. We assume that for some reason unconnected to where the unit has arrived, data about the subsequent development of the unit has been lost. For example, records concerning a group of people after a particular point in time might have been destroyed or simply not recorded because the person doing the recording became ill. The formal term for this type of data is *missing completely at random.*

In this setting we obtain a modification of a complete random sampling scheme called *ancestral sampling.* It is straightforward to check that here the likelihood still takes the same product form given in (5.1). The only difference to complete sampling is that a unit contributes to the counts along an edge only if it both arrives at that situation *and* its subsequent development is recorded. The formula for the posterior distributions of the stage probabilities then as before just counts the number of units arriving at a stage and proceeding along an edge. Compared to random sampling this type of sampling gives even less information about units closer to the leaves than does complete random sampling. It also destroys the property of mass conservation.

Sample survey data

Sometimes surveys are stratified as in Section 2.1.1. Suppose that such a stratification takes place over a cut in a CEG. This might happen for instance when a fixed number of units is sampled at each position defining that cut. Then such counts will tell us nothing about how probable it was for a unit to arrive at a position in the cut. However they will still be informative about the probabilities of each stage each unit traverses after the cut for as long as the unit its observed. This will again give us a likelihood that separates on the stage probabilities and the analogous conjugate Dirichlet analysis: see Exercise 5.1. This type of sampling is widespread in social, medical and public health studies. Especially when sampling is costly in time or money and certain positions will be of more critical interest than others it is often worth stratifying the sampling to sample more units from the categories of interest. Through these types of sampling schemes we can guarantee that we can learn direct resources so as to learn more about the most critical elements of the model. This in particular can help compensate for any degradation of information about the often finer distinctions expressed through positions placed further up a tree.

Experimental data

Often data is available which arises from designed experiments undertaken in laboratories. For instance, designed to investigate the effect at various levels of concentration of a drug which shall alleviate headaches. Usually, a possibly stratified unit is randomly assigned to one of a set of prespecified concentrations of the drug and the effect on the perceived reduction in head pain after a fixed period of time is recorded. To accommodate such experimental data appropriately requires care because the proportion of patients experiencing the effectiveness at a given level of concentration in the trial may be different from the proportion of people in the population of interest that might experience this effectiveness when they take the drug in the real world. However, if it is possible to associate these two proportions—and so their respective edge probabilities in the CEG—then again we obtain a likelihood that separates on the stage parameters. The sorts of hypotheses that allow us to identify two such probabilities are often called *causal hypotheses* and are discussed in Chapter 8.

Combinations of different data sources

It is straightforward to check that the likelihood of a composite of different independent data sources that all give the likelihoods which individually separate gives rise to a composite likelihood that also separates. This is because in this case the composite likelihood is simply the product of the component likelihoods. This simple observation is actually very powerful and one reason we would recommend the use of Bayesian methods (or observed likelihood methods) for inference on CEGs. It not only means that we can combine many different sources of data but also data owned by different people who have particular expertise concerning different parts of the system: see e.g. [7, 99].

An aside: Loss of mass conservation

It is important to notice that although many forms of sampling like those discussed above preserve the closure to distributional assumptions making inference straightforward, under such sampling mass conservation is usually destroyed. We will see in the next section that this can make model selection processes more complicated to enact.

Sampling when data is not missing at random or design

One of the great advantages of the CEG is that it is able to cope with problems where missingness in data is not at random. This is not the case in BN modelling where an almost ubiquitous assumption when addressing missingness in some components in a dataset is that the reason for this missingness is completely unconnected with the process. However in many problems this assumption is transparently false. For example, in a medical or public health setting some readings are more likely to be missing if the nature of the illness would make it pointless to take the measurement. In both this case the absence or presence of a reading is informative about the state of health of the patient. Similarly it is well known that families within which child abuse occurs often move regularly to avoid local authority detection. In fact criminals of all hues tend to try to hide evidence of their activities. As a consequence, missingness not at random is very common in analyses of criminality or indeed to any competitive environments where sharing information is not beneficial to all parties.

Within the context of discrete systems the CEG is a very powerful framework within which to address these issues in a simple and elegant way. In particular, whether or not data is missing in a unit at certain points in its development can simply be introduced as a situation, usually early in the tree underlying the corresponding CEG, splitting the population who exhibit the suspected significant missingness from those who do not. It can be shown that missingness at random then must imply that various subsequent positions—some subsequent to the 'missing' edge category and some to the 'present' category—be placed in the same stage. The methods we suggest here can then be simply adjusted to ask whether there appears to be some non-randomness present and if it is the extent of the effect of the missingness on likely subsequent measurements. Methods for determining whether these two positions should in fact be placed in the same stage are then described in the next chapter. Details of such a methodology are beyond the scope of this book but are carefully discussed and illustrated in the very accessible work of [6].

Sampling with systematic missing variables

Even when data is missing at random, if observations are missing non ancestrally so that potential explanatory variables remain unobserved in most or all units then even with huge datasets estimation can inevitably turn out to be ambiguous. Technically, these statistical problems occur because various edge probabilities might turn out to be unidentifiable or at least approximately so. By carefully reflecting on these inferential issues the inherent difficulty becomes entirely explicable—the way we have articulated the problem actually *logically* entails this ambiguity. However this does not prevent these ambiguities being introduced by accident. If, through a sloppy analysis, we identify one of the potential explanations of the data but not all the others then our analysis can seriously mislead us.

These problems first came to light in the study of discrete BNs. For example, they are endemic in the study of phylogenetic trees where we try to reconstruct the evolutionary past from the status of the current population [112]. Even the simplest of examples embody these serious ambiguities. Furthermore such ambiguities can be very subtle and may not just be due to for example aliasing problems: see e.g. [71].

Of course because the discrete BN is a particular example of a CEG, CEGs can also give rise to highly multimodal likelihoods on various interior probability parameters where those internal stages are not directly recorded on those units: see Exercise 5.2.

Modelling evolving probabilities and dynamic CEGs

It is not uncommon for the relationship between measures on a population of units to be hypothesised to drift in time. For example, consider yearly cohorts of students studying on various university programmes. The probability distributions of performance of those students in one yearly entry cohort may well be strongly linked to those of the previous cohort. However drifts in the quality of the university environment and culture and the match between the provision of the university courses and the school training will inevitably drift year on year. So even if it is reasonable to assume the CEG itself describing a yearly cohort is the same as the previous year's, it would be wise to assume that the stage probabilities might gradually drift away from their baseline year on year.

There is a very simple way of modelling this drift using a multivariate steady model [96] on the analysis given above. In [34] we prove that by adding a single drift parameter which can be estimated we can formally modify the conjugate Dirichlet analyses given above so that such hypothesised drift is modelled. We find that updating equations then follow those in Section 5.1.1 except that data is discounted depending on its age. This means that the sums of counts of units through different stages and floret edges are simply replaced by exponentially weighted sums of counts. But posterior probability distributions on the stage remain product Dirichlet and so amenable to fast and transparent inference.

Another critical area where development still proceeds at a pace are systems that consist of processes which indefinitely continue to develop. There is now a considerable amount of technologies available for dynamic BNs and these are proving extremely useful in a number of applications [62]. We have recently demonstrated that analogous dynamic CEGs can be developed with similar properties. It has been shown that these can often be thought of as particularly useful classes of semi-Markov processes [3, 70]. Examples of their application included models of emerging behaviours. Again there is no space in this small text to properly develop a full description of these model classes and some of the inferential tools needed for their use are still under development. However, the class of dynamic CEG models was first formally defined and analysed in [4] and more recently in [16] where we developed more practically usable subclasses of these processes.

5.2 Propagating information on trees and CEGs

One problem we face when modelling with CEGs is that, just as with the BN, in practice these structures can be huge. In such a large system, how can we efficiently calculate a conditional probability of an event associated with a particular individual drawn from the population when we know something but not everything about her? Although Bayes' Rule gives us a formula for this conditional probability—see Section 4.1.3—because of the size of the CEG it may be impossible to accurately store within a computer all the minute atomic probabilities in the space in order to use that formula. This issue has been very extensively studied when the CEG is also a BN and methods to address this are called *propagation algorithms*: see Section 2.3.5. Here we first demonstrate that there are exact analogues of BN propagation algorithms as these apply to saturated probability trees. We then report more efficient methods that have been developed to find propagation algorithms applicable to CEGs [108].

A second question is how do we calculate these conditional probabilities, inferred from propagation algorithms, when the population probabilities are themselves uncertain? We show that under very general conditions this question is straightforward to answer in an entirely formal way.

5.2.1 Propagation when probabilities are known

Suppose that we have some information about individual units in a population available. For instance in the example on liver and kidney disorders from Section 5.1.3, we might have learned that a patient has already had some form of surgery. This would correspond to knowing whether or not certain unfoldings in the graph representing this population have already happened or were made impossible by some historic development. Then it would not make sense to reassign the initial clinical treatment. We might instead like to evaluate the patient's future unfoldings conditional on that new knowledge. Below we present how we can tackle this problem for both probability trees and CEG, and we then illustrate our results in an example setting in Section 5.2.2 below.

Propagating information on a saturated probability tree

We begin with a straightforward propagation method that simply forgets all the stage structure of a staged tree—treating any such graph as a saturated tree—and propagates information we receive about a particular unit to provide a valid probability tree conditional on that information. Although this is an inefficient propagation, unlike the BN propagation algorithms it is appropriate for *any* information we have learned about a unit, so any event in the associated path-sigma algebra might be known to have happened. This approach provides an accurate algorithm for producing a new set of probabilities on the given tree avoiding the inevitable rounding error problems associated with more naïve methods: compare the discussion in Section 2.3.5.

Any information we could learn about a unit of interest passing through a probability tree $(\mathcal{T}, \boldsymbol{\theta}_{\mathcal{T}})$ can be represented by a particular subset of root-to-leaf paths $A \subseteq \Lambda(\mathcal{T})$ representing atoms which have occurred and others $\Lambda(\mathcal{T}) \setminus A$ which have not. Following the development that led to Proposition 4.6, we can express the atomic probabilities conditional on A by a straightforward application of Bayes' Rule. Thus

$$p_{\boldsymbol{\theta},\mathcal{T}}(\lambda \mid A) = \begin{cases} P_{\boldsymbol{\theta},\mathcal{T}}(A)^{-1} p_{\boldsymbol{\theta},\mathcal{T}}(\lambda) & \text{if } \lambda \in A \\ 0 & \text{if } \lambda \in \Lambda(\mathcal{T}) \setminus A \end{cases} \tag{5.8}$$

where we use the shorthand notation $P_{\boldsymbol{\theta},\mathcal{T}}(A) = \sum_{\lambda \in \Lambda(\mathcal{T})} p_{\boldsymbol{\theta},\mathcal{T}}(\lambda)$.

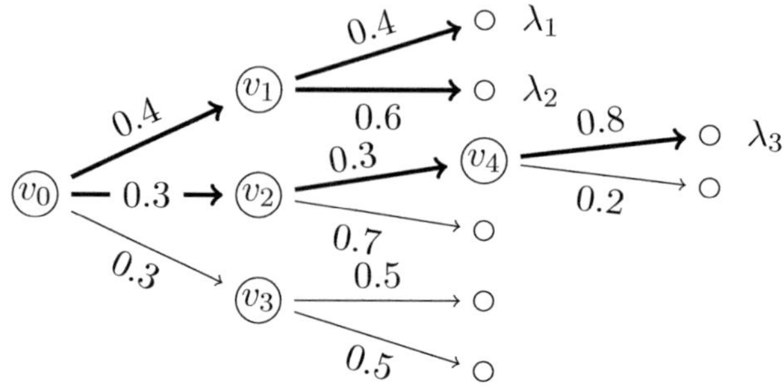

(a) A saturated probability tree $(\mathcal{T}, \boldsymbol{\theta}_{\mathcal{T}})$.

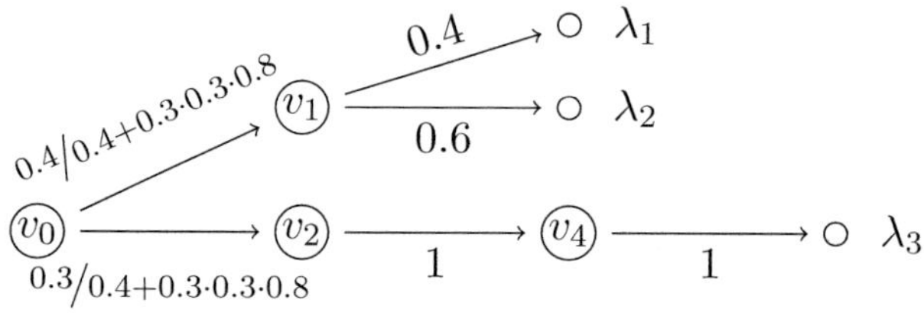

(b) A (degenerate) probability tree with graph $\mathcal{T}_{\neq 0} \subsetneq \mathcal{T}$ after propagation.

FIGURE 5.4
An example of propagating information on a saturated probability tree. Here, $A = \{\lambda_1, \lambda_2, \lambda_3\}$ is the observed event, depicted thick in the graph above.

Now how all the edge probabilities under (5.8) must change in the light of having observed A can be expressed using a sequence of straightforward local arithmetic operations. In practice, we simply update our estimates or beliefs about the system we analyse and adjust the remaining parameters. This is to ensure that the probability mass function associated with the updated system—after conditioning on A—is still strictly positive and sums to one across all florets. We hereby exploit that every atomic probability is the product of labels and that the probability of every vertex-centred event can be interpreted as a weight associated to that vertex. Explicitly, this is how this is done:

1. Attach the number 1 to each situation in $\mathcal{T}$.
2. Set all edge probabilities $\theta(e)$ which do not lie along root-to-leaf paths in A, so for all $e \notin E(\lambda)$ for any $\lambda \in A$, to zero.
3. Calculate the new probabilities $P_{\boldsymbol{\theta},\mathcal{T}}(\Lambda(v))$ for each situation v which is the parent of a leaf and for which the associated vertex centred-event lies in A, so $\Lambda(v) \subseteq A$. There must be at least one such situation. Replace 1 by $P_{\boldsymbol{\theta}}(\Lambda(v))$ for each of these. Label this set of situations as *actuated.*
4. Replace each of the edge probabilities $\theta(e)$ for $e \in E(v)$ by the fraction $P_{\boldsymbol{\theta}}(\Lambda(v))^{-1}\theta(e)$.

5. Multiply all edges leading into an actuated situation by $P_{\boldsymbol{\theta}}(\Lambda(v))$. Now the new edge probabilities associated with the subtree $\mathcal{T}_{\neq 0} \subseteq \mathcal{T}$ which can be constructed from $\mathcal{T}$ by deleting all the edges and their receiving vertices which were assigned probability zero is a valid probability tree with edge probabilities proportional to their true values.
6. Iterate the above process on $\mathcal{T}_{\neq 0}$ and repeat it until there is only a single floret remaining.

In the algorithm above, we propagate information backwards through the tree, first actuating vertices using downstream information and then renormalising the new edge probabilities. In particular, in step 5 we preserve all the probabilities of root-to-leaf paths in A and set all probabilities of root-to-leaf paths in $\Lambda(\mathcal{T}) \setminus A$ to zero. In $\mathcal{T}_{\neq 0}$ we cancel all unfoldings from $\mathcal{T}$ which have zero probability. Then by construction, in step 6 all probabilities in the last floret—which will be the root vertex of $\mathcal{T}$—will have edges that sum to one. When these recalculated edge probabilities are attached to $\mathcal{T}$ then we have calculated all the edge probabilities conditional on A.

This algorithm works for any event $A \subseteq \Lambda(\mathcal{T})$. Because each operation is local and will typically involve summing, multiplying and dividing numbers that are not usually close to zero, we have avoided the problem of expressing probabilities as a large sum of some extremely small probabilities as we would need to do when applying Bayes' Rule naïvely—giving rise to potentially crippling rounding errors that may well have occurred.

The roll-back method presented above turns out to be a surprisingly useful springboard for designing more elaborate algorithms. However in the form given above it uses none of the stage information in the tree and so is unnecessarily inefficient in many cases—enormous numbers of calculations can be duplicated when using this method. Under two conditions can we extend these methods from a staged tree to the CEG framework and make it much more efficient: when the event A is restricted to take a particular form—as outlined below—and provided that the CEG does neither have a huge number of positions nor a position with a huge number of edges. Then we can calculate the revised position probabilities on the CEG, both accurately as in the sense above and also almost instantaneously for most problems even when the underlying event space has a very large number of atoms. We show below that this can be easily done using a 'backward' and a 'forward' step.

Propagating intrinsic information on a CEG

Here we present the algorithm proposed by [108] to propagate information over a CEG whose conditional probability tables are known. This algorithm is analogous to one developed for retraction of evidence in a BN [21].

In general, not all kinds of evidence can be propagated efficiently through a CEG. In a BN framework this is a well-known issue because passing information based on an arbitrary function of the problem variables using local messages can disrupt the conditional independence structures upon which the propagation algorithm relies to perform the local updates. For the methods proposed below, we will need to assume that the information $\mathcal{I}$ we have available about a CEG $\mathcal{C} = (W, E)$ is 'identifiable' in terms of some subset of the sample spaces of the random variables associated with the set of positions W. This simply means that the information can be represented graphically by an intrinsic event: see Section 4.1.2. Formally, we need to assume that there exists a set of edges $E(\mathcal{I}) \subseteq E$ such that $\mathcal{I}$ can be expressed as a set of paths

$$\Lambda(\mathcal{I}) = \{\lambda \in \Lambda(\mathcal{C}) \mid E(\lambda) \subseteq E(\mathcal{I})\}. \tag{5.9}$$

Information $\mathcal{I}$ is said to be *trivially* $\mathcal{C}$-intrinsic if $E(\mathcal{I})$ includes all edges of the CEG, so $E(\mathcal{I}) = E$. Often $E(\mathcal{I})$ is chosen to be as small as possible.

The intrinsic events of a CEG are typically much more varied than their BN counterparts, even if a CEG is also a BN: they are not simply subsets of the sample space of predetermined variables but *events* in that space. Compare also the development in Section 4.1.

Propagation of new information using a BN model requires us to perform a pre-processing step. In this step, the corresponding DAG representation needs to be moralised, triangulated and transformed into a junction tree whose vertices are cliques of variables: see Section 2.3.5. This enables us to identify the most relevant conditional independence structures embedded into a model and to translate them into an appropriate form for efficient probability propagation. In a similar way, the CEG has to be prepared for propagation. Here, the pre-processing is rather simple and consists only in removing the colours of the initial CEG graph: so rather than forgetting about the stage structure of the underlying staged tree as in the first algorithm presented above, we here forget about the stages of a CEG, so only those vertices in the staged tree which are in the same stage but not in the same position. The resulting simple CEG $\mathcal{C}_t$ is called the *transporter* of $\mathcal{C}$. So by construction, both $\mathcal{C}_t$ and $\mathcal{C}$ have the same graph except that $\mathcal{C}_t$ can be regarded as uncoloured. As a consequence, the positions now play a role similar to that of cliques in the BN framework.

Given an intrinsic information in a CEG $\mathcal{C}$ and a transporter CEG $\mathcal{C}_t$, we can update our belief using a two-step algorithm we describe below.

Step 1: Evidence collection

Suppose we are given information $\mathcal{I}$ about a CEG $\mathcal{C} = (W, E)$ that can be represented using a set of root-to-sink paths $\Lambda(\mathcal{I}) \subseteq \Lambda(\mathcal{C})$ with edges $E(\mathcal{I})$ as in (5.9). We index the sets of positions and edges in $\mathcal{C}$ such that if $i < j$ then neither w_i nor e_i lie downstream, respectively, of w_j and e_j along any root-to-sink path in $\mathcal{C}$, for $w_i, w_j \in W$ and $e_i, e_j \in E$. Let further $K+1 = \#W$ be the total number of positions and let $K_i = \#E(w_i)$ denote the number of edges of each position, $i = 1, \ldots, K$. Replace the CEG $\mathcal{C}$ by its transporter CEG $\mathcal{C}_t$. To simplify notation, let still $\Lambda(\mathcal{I}) \subseteq \Lambda(\mathcal{C}_t)$.

In this first step each position absorbs the new information $\mathcal{I}$ using a *backward* strategy: propagating from the sink w_∞ backwards to the root w_0. Hereby, the collected evidence is stored for each position w_i using a pair $(\boldsymbol{\tau}(w_i), \phi(w_i))$ where the first entry $\boldsymbol{\tau}(w_i) = (\tau_1(w_i), \ldots, \tau_{K_i}(w_i))$ is a vector whose component $\tau_j(w_i)$ is a *potential* corresponding to propagated evidence of arriving (backwards) at w_i through an outgoing edge e_{ij} for $j = 1, \ldots, K_i$; and the second entry $\phi(\omega_i)$ is an *emphasis* that merges additively all potentials arriving at w_i. Explicitly,

$$\tau_j(w_i) = \begin{cases} 0 & \text{if } e_{ij} \notin \Lambda(\mathcal{I}), \\ \theta(e_{ij}) & \text{if } e_{ij} \in \Lambda(\mathcal{I}) \end{cases} \tag{5.10}$$

where $\theta(e_{ij})$ denotes the label of the edge e_{ij}, $j = 1, \ldots, K_i$, and

$$\phi(w_i) = \sum_{j=1}^{K_i} \tau_j(w_i) \tag{5.11}$$

for all $i = 1, \ldots, K$.

A position w_i is said to be *accommodated* whenever its pair $(\boldsymbol{\tau}(w_i), \phi(w_i))$ has been calculated. In this step all positions should be accommodated in its reverse order excluding the sink position w_∞ which we will denote $w_\infty = w_K$; so from w_{K-1} to w_0.

Step 2: Evidence distribution

In the second step all collected evidence is distributed *forwards* in the accommodated transporter CEG $\mathcal{C}_{\mathrm{t}}$. This operation uses the potential and emphasis of each position to deliver a revised vector of floret probabilities $\hat{\boldsymbol{\theta}} = (\hat{\boldsymbol{\theta}}_0, \ldots, \hat{\boldsymbol{\theta}}_{K-1})$ conditional on the information $\mathcal{I}$. Explicitly, for all $w_i \in W$, $i = 1, \ldots, K-1$, we set:

$$\hat{\theta}(e_{ij}) = \frac{\tau_j(w_i)}{\phi(w_i)} \qquad \text{for } j = 1, \ldots, K_i. \tag{5.12}$$

Denote by $\hat{\mathcal{C}}_{\mathrm{t}} = (\mathcal{C}_{\mathrm{t}}, \hat{\boldsymbol{\theta}}_{\mathcal{T}})$ the CEG which has the same graph as the transporter $\mathcal{C}_{\mathrm{t}} = (\mathcal{C}_{\mathrm{t}}, \boldsymbol{\theta}_{\mathcal{T}})$ but edge probabilities $\hat{\boldsymbol{\theta}}_{\mathcal{T}}$ updated with respect to the original underlying staged tree. For any position w in the resulting CEG $\hat{\mathcal{C}}_{\mathrm{t}}$ the updated probability of arriving at w along a root-to-w path which we simply denote $w_0 \to w$ can thus simply be calculated as in any CEG with a probability distribution:

$$P_{\hat{\boldsymbol{\theta}},\mathcal{T}}(w_0 \to w \mid \Lambda(\mathcal{I})) = \prod_{e \in E(w_0 \to w)} \hat{\theta}(e). \tag{5.13}$$

It therefore follows directly that the updated probability to arrive at position

w has the form

$$P_{\boldsymbol{\theta},\mathcal{T}}(\Lambda(w) \mid \Lambda(\mathcal{I})) = \sum_{\lambda \in \Lambda(w)} P_{\boldsymbol{\theta},\mathcal{T}}(w_0 \to w \mid \Lambda(\mathcal{I})) \tag{5.14}$$

in the notation of Proposition 4.6.

In analogy to the BN propagation algorithm [21, Equation 6], each atom represented by a w_0-to-w_∞ path $\lambda \in \Lambda(w_\infty) = \Lambda(\mathcal{C}_\mathrm{t})$, has therefore an attached atomic probability which can be expressed in terms of potentials and emphases:

$$\pi_{\hat{\boldsymbol{\theta}},\mathcal{T}}(\lambda) = \prod_{e_{ij} \in E(\lambda)} \frac{\tau_j(w_i)}{\phi(w_i)} \tag{5.15}$$

by (5.12) to (5.14). From (5.15) we can see that the computational cost of the propagation algorithm can be substantially reduced if a smaller CEG whose vertices and edges are just those with non-zero emphasis and potential in $\hat{\mathcal{C}}_\mathrm{t}$ is used. This can be easily done because any non-trivial $\mathcal{C}$-intrinsic information strictly enables us to reduce the number of edges in $\hat{\mathcal{C}}_\mathrm{t}$.

We illustrate the use of this algorithm in an outworked example in the next section.

Pseudo-code for the propagation algorithm

We can now present pseudo-code for the propagation algorithm we developed above. Although shown here in a modified version, the proof of this algorithm is identical to the one initially introduced in [108]. Here, the order in which the computations are carried out is completely defined by the graph of the original transporter CEG $\mathcal{C}_\mathrm{t}$ and so can be set beforehand.

The reduced CEG $\hat{\mathcal{C}}$ resulting from this algorithm is not necessarily minimal in the sense that there might exist two different vertices which are in the same position under the updated probabilities $\hat{\boldsymbol{\theta}}_w = \hat{\boldsymbol{\theta}}_{w'}$ for $w \neq w'$ but which are not merged in the graph. However, this graph can of course subsequently be coloured according to these updated probabilities. Clearly it is then possible to obtain a minimal coloured CEG if a third algorithm step is included for this purpose. This however is not strictly necessary for evidence propagation and the interpretation of the results.

Algorithm 1: The propagation algorithm

Input: A CEG $\mathcal{C} = (W, E)$ whose positions are numbered with probability mass function p and information $\mathcal{I}$ represented by an intrinsic event.

Output: An uncoloured CEG $\hat{\mathcal{C}}_{\mathrm{t}} = (\hat{W}, \hat{E})$ with updated probability mass function $\hat{p}$.

Set $\hat{W} = \emptyset$, $\hat{E} = \emptyset$, and $\phi = 0$.
Initialise a vector $\hat{\boldsymbol{\theta}}$ of length $\#W - 1$.
for i *from* $\#W - 1$ *to* 0 **do**
    Initialise a vector $\boldsymbol{\tau}_i = \mathbf{-1}$ of length K_i
    for j *from* 1 *to* K_i **do**
        $\tau_{ij} \leftarrow \tau_j(w_i)$ from (5.10)
        if $\tau_j(w_i) \neq 0$ **then**
            $\hat{E} \leftarrow \hat{E} \cup \{e_{ij}\}$
    $\phi \leftarrow \phi(w_i)$ from (5.11)
    if $\phi(w_i) \neq 0$ **then**
        $\hat{W} \leftarrow \hat{W} \cup \{w_i\}$
    $\hat{\boldsymbol{\theta}}_i \leftarrow {}^{\tau}/_{\phi}$
return $\hat{W}, \hat{E}, \hat{\boldsymbol{\theta}}$

5.2.2 Example: Propagation for liver and kidney disorders

Return to the example analysed in Section 5.1.3. Suppose that a patient who has worked overseas has had a liver and kidney disorder during the past year. Returning to his home country, he visits a specialised clinic and he reports to a physician that his case was diagnosed as non-critical but that a semi-elective surgery was prescribed to him. Before requiring further clinical exams, the physician would like to assess his actual state of health using the data collected during the clinical interview. For this purpose, he uses a decision-making support software based on the CEG given in Figure 5.1b whose conditional probabilities were fixed to have the posterior means given in Table 5.1.

In this case, the information $\mathcal{I}$ available about the patient can be represented by the set of edges

$$E(\mathcal{I}) = \{(w_0, w_1), (w_1, w_4), (w_1, w_4), (w_2, w_4), (w_4, w_\infty, \mathrm{M}), (w_4, w_\infty, \mathrm{LT})\}$$

where, if otherwise ambiguous, the third entry in an edge denotes the label of that edge. After introducing the information $\mathcal{I}$ into the software, the physician obtains the updated CEG $\hat{\mathcal{C}}_{\mathrm{t}}$ in Figure 5.5b. Figure 5.5a shows the transporter CEG $\mathcal{C}_{\mathrm{t}}$ that was used to propagate the information with the corresponding potentials and emphases in order to obtain the updated CEG. For this patient, the odds between a minor or a serious disorder are then $2:1$ whilst the odds between monitoring and lifetime medication are $5:3$.

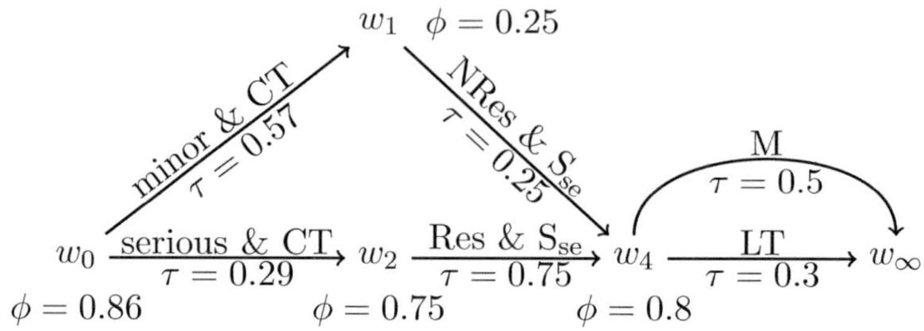

(a) The transporter CEG $\mathcal{C}_t$ with a potential τ associated to each edge and an emphasis ϕ associated to each vertex.

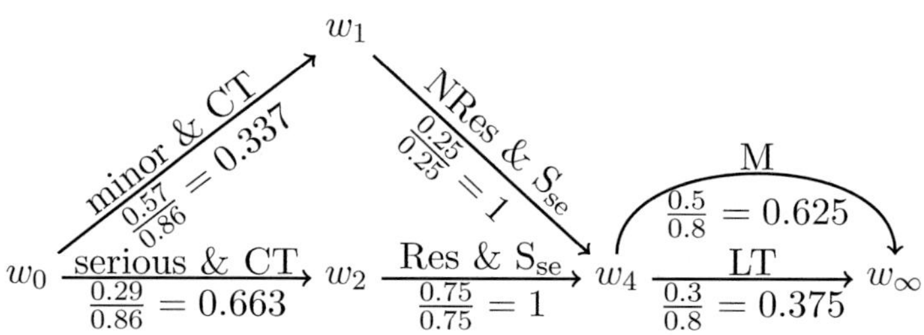

(b) The updated CEG $\hat{\mathcal{C}}_t$. Each edge is now labelled by an updated non-zero probability.

FIGURE 5.5
The transporter and updated CEGs associated with the CEG depicted in Figure 5.1b assuming a patient is known to be in a non-critical state and had been submitted to semi-elective surgery.

Here, the conditional probabilities associated with the updated CEG can be stored using only 14 cells. This is much more efficient than the standard approach of representing and propagating this information using a BN model. This is not only because of the context-specific information but also because the process depicted in Figure 5.1a develops in a highly asymmetric way. To see this consider a possible 5-variable BN to model this process as depicted in Figure 5.6 and where the random variables are as follows: X_1 denotes the category of dysfunction with states *minor*, *serious* and *critical*; X_2 indicates whether a patient responds to the clinical treatment; X_3 denotes the possible types of surgery with states *none*, *semi-elective* and *emergency*; X_4 flags whether or not a patient is alive, and X_5 denotes the final response with states *recovery*, *monitoring*, *medication* and *death*.

The responses to the first clinical treatment which were stored in the set of positions $\{w_1, w_2, w_3\}$ in the CEG are now represented through the variables X_2, X_3 and X_5. The variable X_5 also stores the information on the result of a surgery, which corresponds to the set of positions $\{w_1, w_4, w_6\}$ in the CEG.

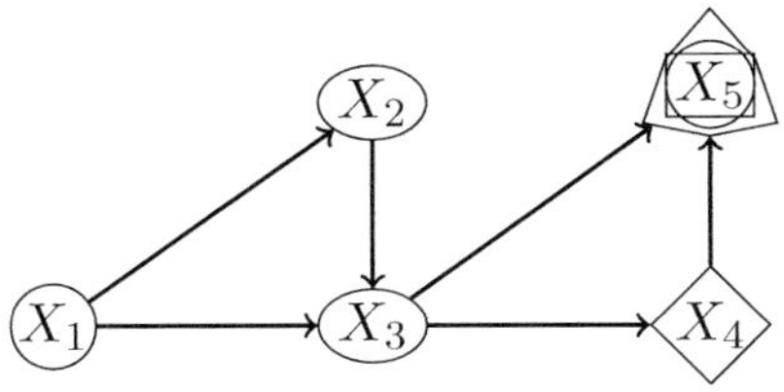

FIGURE 5.6
A DAG for the liver and kidney disorder BN model. The shape of each vertex corresponds to the shape of the positions asssociated with the same meaning in the CEG from Figure 5.1b.

The BN model will thus require 42 cells—18 for the clique $\{X_1, X_2, X_3\}$ and 24 for the clique $\{X_3, X_4, X_5\}$—to store these conditional probabilities. Of these, 28 store the value zero—12 in the clique $\{X_1, X_2, X_3\}$ and 16 in the clique $\{X_3, X_4, X_5\}$. This results in a greater computational cost for propagating evidences than when using a CEG model which provides us with a much more parsimonious storing structure.

The benefits of using CEGs for propagation as seen in this example translate to much more general settings. Particularly in non-product-space settings is the propagation of evidence using CEGs often much more efficient than analogous computations using BNs. This is because a CEG graph can embody such asymmetry by having root-to-sink paths of very different lengths. It can thus provide a framework enabling us to avoid performing repetitive calculations and propagating zeros in sparse but large probability tables; for example, were we to use the simple base algorithm discussed at the beginning of this section. As we have demonstrated in the example setting above, this would not be possible if we used a BN model with a naïve propagation algorithm. Moreover, the range of intrinsic information with a CEG model is larger and more general than those amenable when using a BN model. This allows us to perform more detailed analysis and use a finer information partition which is especially important when data is heterogeneously censored.

5.2.3 Propagation when probabilities are estimated

In the development of propagation methods in the previous section we have exclusively considered the case where a CEG's edge probabilities were given. But what happens when we have chosen to estimate these from data as in the prior-to-posterior analysis of Section 5.1? In order to answer this question again it is simplest first to consider the saturated model which can without loss be represented by a star tree and to then generalise from this case.

Consider a saturated staged tree $(\mathcal{T}, \boldsymbol{\theta}_{\mathcal{T}})$ represented by a single floret $\mathcal{T} = (v_0, E)$ with $n = \#E$ leaves. Suppose we have placed a Dirichlet distribution $\text{Dir}(\boldsymbol{\alpha})$ on the vector of edge probabilities $\boldsymbol{\theta}_{\mathcal{T}}$. Then from the results we illustrated in Section 2.1 we see that conditional on any set $A \subseteq E$ the distribution on the vector $\boldsymbol{\theta}_A = (\theta(e) \mid e \in A)$ of edge probabilities associated with leaves $\lambda \in A$ has a joint Dirichlet distribution $\text{Dir}(\boldsymbol{\alpha}_A)$ where again $\boldsymbol{\alpha}_A$ denotes the subvector of $\boldsymbol{\alpha}$ whose indicies lie in A. Of course the probability of any $\lambda \notin A$ is zero with probability one. As a consequence, we here immediately obtain the full joint distribution of the new edge probabilities.

There are two points worth mentioning. The first is that when we write this update in terms of the distribution of $\boldsymbol{\theta}_{\mathcal{T}}$ prior to observing A on the individual of interest and compare this to the posterior on $\boldsymbol{\theta}_A$, written in terms of the expectations and effective sample size parameters, after substitution we obtain the following. First, the posterior mean vector $\boldsymbol{\mu}_{A+}$ of the probabilities $\boldsymbol{\theta}_A$ is given by

$$\mu_{\hat{A}+} = \frac{\mu_{A+}}{\overline{\mu_A}} \tag{5.16}$$

where $\overline{\mu_A} = \sum_{i \in A} \mu_i$. We can thus see that the posterior mean of these probabilities satisfies the same formula to calculate the posterior mean $\boldsymbol{\mu}_{A+}$ as the one we would use on $\boldsymbol{\theta}_A$ were we to know the vector of edge probabilities. Second, the effective sample size reduces from $\overline{\alpha} = \sum_{i=1}^{n} \alpha_i$ to $\alpha_A = \sum_{j \in A} \alpha_j$. Thus whilst we may have learned for certain that a person of interest has a profile consistent with A, typically the variance on the probabilities conditional on this event increases when the event becomes smaller. This increase is particularly extreme if the category we observe in our unit of interest is unexpected so that a prior α_A is small relative to α. This is because typically there will have been fewer people in our real and phantom sample matching the profile of the person in front of us. Without more structural information enabling us to match the person of interest with aspects of many others in the population partially matching him then of course we can learn very little more that distinguishes different further refinements consistent with what we know about him as learned from A. On the other hand this emphasises precisely why it is so valuable to encode any further structural information—for example via a CEG—so that we can estimate conditional probabilities more accurately from the population.

Next we turn to the general case of a CEG $\mathcal{C}$ where we want to learn the uncertain distribution after observing an intrinsic event $A \subseteq \Lambda(\mathcal{C})$ concerning a given unit. The marginal distribution of that unit has each of its atomic probabilities given by the products of the conditional means along the corresponding root-to-sink path. Provided the unit has been drawn randomly from the population, we can simply substitute the values of these conditional means into our propagation scheme given in the previous section. This then gives us the expectation of the posterior probability mass function of the unit of interest.

For the general CEG with its conjugate prior, when we observe an intrinsic event conjugacy is preserved in the sense given below. This is because any intrinsic event A can then be expressed as the intersection of a number of events associated with a particular stage u_i with probability vector $\boldsymbol{\theta}_i$ and with a Dir($\boldsymbol{\alpha}_i$) distribution: compare Proposition 4.1. Hence, as in the saturated example above, conditioning on A and taking the inference stage by stage, the subvector $\boldsymbol{\theta}_{i,A}$ of stage probabilities not zeroed by the conditioning in this graph has a Dir($\boldsymbol{\alpha}_{i,A}$) distribution—with all other edges in the floret being assigned a zero probability with probability one.

It thus follows that the non-zeroed atomic probabilities will have a new product Dirichlet distribution, defined by the subCEG of possible developments after A with the posterior distributions calculated above. It is important to note that when the stage structure is sufficiently rich this will demonstrate that we can learn these probabilities with much more confidence than for the saturated model. For example Exercise 5.3 proves that the bounds on the variance are much tighter here than in the saturated case. The conditional effective sample sizes of the conditioned stages, though smaller than the sizes of those in the full joint mass function, will typically be moderately large. So again the variance of the conditional probabilities we need will be small and we will be a lot more sure of these estimates.

When the probability mass function of the containing population is uncertain the propagation schemes of the posterior edge mean probabilities on the unit by conjugacy are independent of each other. But it is also possible to develop tower rules to calculate the variances of the atomic conditional probabilities associated with a unit of interest. It can be shown that these must always apply provided that the vectors of different posterior stage probabilities are mutually independent whether or not we have conjugacy. For analogues of these recurrences as they apply to BNs see [22].

Because learning on CEGs with only a moderate number of stages is much faster than their saturated analogue the variances of the floret probability vectors tend to be orders of magnitude smaller. It follows that when we learn about the population we can also accurately learn about new individuals drawn from the estimated population for whom we have a partial classification. Typically the only time when this might not be so is if the person being seen exhibits combinations of features which were a priori very surprising: see the discussion for the saturated case above. But then surely we should be wary and we might not have yet given much prior consideration to cases like the one in front of us. We might then like to glean more information before proceeding to act.

5.2.4 Some final comments

We have seen in this chapter how to estimate edge probabilities in a given Chain Event Graph and how to update these in light of new information.

In general, when we learn about a particular CEG from data as with any statistical model, it is always important to check that its hypothesised structure is broadly consistent with the data we have observed: so for instance that the stage structure and its implications in terms of conditional independences between depicted events makes sense for the population we model. There are many well-studied ways of doing this that can be borrowed from standard statistical texts. For example, we can always test to see whether various situations are in the same stage by using standard checks on sets of Multinomial samples having the same distribution. More consistently with the Bayesian methodology used in this book we can instead use Bayes Factor methods to do this—something we will do implicitly if we conduct any model search—as explained in the next chapter.

Many other methods used for BNs [22] can be simply adjusted to provide CEG analogues. The prequential methods described there are especially easy to generalise to this context. More subtly it is wise also to check that the samples we use really do appear to be random. Sadly there is no space to discuss these important issues here so some of these will be reported in a future publication.

5.3 Exercises

Exercise 5.1. *Suppose in the CEG depicted in Figure 5.1b you observe a random sample of patients suffering from a critical condition. Write down the likelihood of this sample and show that it separates over the stages of this CEG.*

Exercise 5.2. *Assume that* $g_j(\boldsymbol{\theta}_i)$ *is a Dirichlet density* $\mathrm{Dir}(\boldsymbol{\alpha}_i)$ *and let constants* $\rho_{ij} > 0$ *be such that* $\sum_{j=1}^{l} \rho_{ij} = 1$, *for all* $j = 1, 2, \ldots, l$ *and* $i = 1, \ldots, K$. *Prove that if the prior of* $\boldsymbol{\theta}_i$ *for each stage* i *in a CEG has a mixture of* l *Dirichlet densities, so*

$$f(\boldsymbol{\theta}_i) = \sum_{j=1}^{l} \rho_{ij} g_j(\boldsymbol{\theta}_i) \quad \text{for all } i$$

then the posterior density also has this form. Calculate this explicitly.

Exercise 5.3. *Prove that the variance of an atomic probability* $p_i = \prod_{j=1}^{s} \theta_{ij}$ *associated with a root-to-leaf path going through* s *different stages* $u_{i1}, u_{i2}, \ldots, u_{is}$ *with respective effective sample sizes* $\alpha_{i1}, \alpha_{i2}, \ldots, \alpha_{is}$ *will have a variance bounded above by*

$$(\alpha_{i1} \cdot \alpha_{i2} \cdot \ldots \cdot \alpha_{ik})^{-1} p_i$$

for any $i = 1, \ldots, n$.

Consider the prior distribution of the CEG examined in Section 5.2.2. Calculate these bounds after observing the event that the person in front of you is suffering a critical condition. Find an explicit formula for the variance of w_6 *in Figure 5.1b. Compare this to the bound above.*

Exercise 5.4. *Redo the propagation provided in Section 5.2.2 for the particular case that we observe a patient with a non-minor condition who does not respond to treatment.*

Exercise 5.5. *In the forensic example from Section 1.2 assume you have learned that nobody shared the DNA of the suspect. Propagate this information through the staged tree in Figure 1.1.*

6

Model selection for CEGs

In the previous chapter we showed how when given a CEG we can perform a full Bayesian analysis. However it is commonly the case that whilst we might have various conjectures about a process of interest, we might also want to entertain the possibility that a model describing this process could a priori be any candidate CEG model on the same atoms. In this case, we will not want to commit to a CEG representation until we have first checked this against an available data base. In other circumstances, we might have a conviction that an event tree over a given set of atoms will be appropriate but we might be unclear which colouring that graph should have. We will then usually perform an exploratory data analysis of a random sample of the population of interest and find the CEG that fit this data best. This will put us in the realm of model selection over classes of CEGs which is the topic of this chapter.

Now in principle from a Bayesian perspective this model selection should be straightforward provided that we believe that the data-generating mechanism is indeed drawn from the class of CEGs we search. If this is the case then as illustrated in Section 2.1.4, we simply assign a prior probability to each CEG in terms of its tree and its staging, perhaps a collection of independent Dirichlet distributions. We can then calculate a (log) marginal likelihood in closed form for each of the competing models. Multiplying the prior probability by this marginal likelihood calculated from the data, we obtain a number proportional to the posterior probability of each CEG. In this way we can identify those models which are a posteriori most probable and in this sense describe our data best.

The practical implementation of this MAP method is fraught with difficulties:

1. We have seen in Section 5.1.2 that setting the prior distribution even for a single CEG often needs a considerable amount of care. If we have millions of candidate models within our selection class, how could we possibly elicit all these distributions individually?

2. The different types of stagings of a given CEG increase super exponentially with the number of situations. As a consequence, even if we are able to set up hyperparameters for each model and even when each model in a chosen class can be scored in closed form then searching the space for good explanatory models remains a significant challenge.

3. When a dataset concerns millions of units then vanilla Bayes Factor methods as in Section 2.1.4 tend to assign higher posterior probability than they should to models that contain less structure than the true data generating process. This means that if we employ these to choose a MAP model then that CEG will tend to have too many stages. But the more stages, the more complex and confusing the underlying explanation.

We have recently shown that these issues can be addressed, and successful search algorithms for various classes of CEGs are now set up.

In Section 6.1 we begin with discussing how to effectively set up the priors on the hyperparameters of a possibly huge set of competing CEGs whose efficacy in explaining what we have seen in a random sample we would like to compare. This addresses (1.) above.

In Section 6.2 we then discuss Bayes Factors as they apply to CEG model selection, and in Section 6.3 we present a recent method based on *dynamic programming* developed by [23] which is able to systematically and efficiently search the full space of stratified CEGs to find the MAP CEG in this class strategies. In Section 3.3 we showed that the class of discrete BNs was a subclass of the class of SCEGs, so we show here in particular that ideas developed for BN model search can be translated so that they apply to SCEGs. Despite the efficiency of dynamic programming, even this search method becomes infeasible as the number of explanatory variables increases. As both the authors of [23, 92] recognised, in more complex cases *greedy search* algorithms which search only over promising subsets of the full search space would then need to be employed. Such algorithms are widely used in regression models. One heuristic approach is to use an adaptation of the well known *Agglomerative Hierarchical Clustering algorithm* tailored to CEG search. In contrast to the dynamic programming algorithm above, this can be applied to *any* CEG. It is also orders of magnitude faster. Although this method can miss the true optimal model our practical experiences with using it on moderate sized problems and simulations is that it rarely produces a solution whose score is not close to that of the generating model. It further seems to compare favourably in simulation tests against some of its popular greedy search competitors [92]. This is therefore currently our greedy search method of choice. These developments address the points (2.) and (3.) above.

The methodologies here are still in their infancy and we expect faster and more scalable methods to the ones we describe to appear in the near future. However, we are now at a stage to be able to assert that at least for some classes of CEGs these methods are viable and have in fact been successfully applied to quite a range of problems, in public health, social processes, and marketing. Many of the algorithms we present in this chapter are now freely available in the `R`-package `ceg` [18] for the practitioner to apply to their own datasets. The development in this chapter will be illustrated in a few toy examples but a full application of all of these methods to a real-world problem will then bring these methods to life in Chapter 7.

6.1 Calibrated priors over classes of CEGs

In Section 5.1 we have shown how to set up a prior distribution over the hyperparameters of a single CEG. We now present how to set up compatible hyperparameters across different CEGs.

Typically when eliciting a model we find that there is often a natural choice of CEG model on a given set of atoms Ω to calibrate other models on this same sample space against. We denote the CEG representing this model by $\mathcal{C}_0$. One default choice—and one we use here—could be a saturated model with no underlying structure at all. At other times there is a current model of choice or one that we have initially elicited.

Suppose that we have this single model available and suitably embellished with a product Dirichlet prior which conserves mass, chosen in the way we described in the previous chapter. Because of the mass conservation property of this prior we can then unambiguously define the number of phantom units α_ω that have arrived at each of the atoms $\omega \in \Omega$ of the space being modelled. The trick we explain below is now to use the mass conservation property and stage independence to reverse engineer the priors of any other CEG $\mathcal{C}$ on this space Ω so that they agree on $\{\alpha_\omega \mid \omega \in \Omega\}$. By doing this we will end up specifying prior beliefs that are close as possible to those of our reference CEG $\mathcal{C}_0$ in a sense we will make clear below. This will ensure model selection is as fair as possible and a model not chosen simply because its prior distribution on its hyperparameters automatically lead to a model which fits the data better.

This prior calibration of all the candidate models is extremely simple to achieve provided the class of CEGs is square free: a very mild condition as discussed in Section 3.4. To obtain the value of a stage's Dirichlet hyperparameters $\boldsymbol{\alpha}_i = (\alpha_{i1}, \alpha_{i2}, \ldots, \alpha_{iK_i})$, for stages numbered $i = 1, 2, \ldots, K$, of a CEG $\mathcal{C}$ with K stages, we simply choose $\alpha_{ij} = \sum_{\omega \in \Omega_{ij}} \alpha_\omega$ where $\Omega_{ij} \subseteq \Omega$ is the set of atoms represented by root-to-sink paths in the CEG arriving at a stage w_i and then passing along the edge labeled j, for $j = 1, 2, \ldots, K_i$ and $i = 1, 2, \ldots, K$. This notation has been introduced at the beginning of Chapter 5. Because the prior settings of each alternative model is thus expressed as a function of just one elicitation there is no extra elicitation needed before model selection begins. The only additional prior selection needed is to determine which structural subclass of square-free CEGs we plan to search over: that is which underlying narratives might make sense within the modeled domain. Furthermore because the relationships between the different priors is just a sum, all priors can be calculated quickly even for huge numbers of candidate models. All the routine selection priors we use over the rest of this chapter and the next assume we have set up priors over the hyperparameters of different models in this way. Compare also the discussion in Section 7.5.3.

This method of setting up priors applies to any CEG model subclass of

$\boldsymbol{C}(\Omega)$, regardless of its underlying event tree representations. This is important for sometimes a class of interest will entail a fixed tree, but varying stage structures but in other scenarios we need to contemplate different underlying explanatory trees. In either case the only constraint is that we need any considered CEG to be square-free.

There are some very good reasons for adopting the strategy described above. Many of the properties demanded by, for example, Bayesian BN model selection procedures can be shown to be satisfied through this protocol, as we explain below. It also usually has enough flexibility to be useful for selecting models in quite complex spaces of CEGs.

For any of the statistically equivalent CEG representations of the saturated model this construction of the stage floret probability vectors always gives an equivalent prior, in the sense that these will yield the same posterior. In the saturated case, the family of Dirichlet priors is the only type of distribution for which this will be true: see Exercise 6.2. It follows that all representations of the saturated model have the same scores regardless of the data we see. This is a very useful property since in general we would like models that are statistically equivalent and so which make equivalent structural assumptions to be scored the same, as discussed in Section 4.2. Furthermore, the joint Dirichlet distribution is the only non-degenerate family of prior distributions on the atomic probabilities that will exhibit the required independences entailed by the setting demanded. So even the convenient distributional assumptions can be justified from this point of view [33].

It can be further shown, by studying the effects of the swap and resize operators introduced in Section 4.2, that in fact any two statistically equivalent square-free CEGs will always have the same score when we use the priors above, regardless of the data observed. So the pleasing score equivalence property of the saturated model extends much more generally.

It is now easy to check that the difference in BF score functions (2.17) between one model and a neighbouring model—formed from the first by combining two of the stages of a CEG representation together into a single stage—is only affected by the scores associated with the changes made and not scores associated with other stages: these cancel by construction as will be shown in Exercise 6.3. This property is obviously a desirable one since model selection between two competitors focuses only on the aspects of the structural model where they disagree. This is computationally useful since it enables us to search the space incrementally and so much faster. The implications of this result are investigated in the next section.

6.2 Log-posterior Bayes Factor scores

Suppose throughout this chapter that we have a fixed set of atomic events Ω. To perform model selection in CEGs, it is necessary to first define a family of event trees that spans our pre-specified CEG model space $\boldsymbol{C} \subseteq \boldsymbol{C}(\Omega)$ over these atoms. This will sometimes be a single tree but can be many trees depending on our approach, as discussed over the next section. We also need to specify the stage structures allowed in the chosen class: for instance one of the classes of CEGs presented in Section 3.4. These two elements will constitute the model search space $\boldsymbol{C} = \{\boldsymbol{M}_{\mathcal{C}} \mid \mathcal{C} \text{ is a CEG of the chosen type}\}$. Rather than searching over the models $\boldsymbol{M}_{\mathcal{C}} \in \boldsymbol{C}$, we will search instead over the graphs directly, so over all CEGs representing these models.

We next need to specify a score that will be used to compare any two models within such a search space. There are a variety of methods available. Here we again adopt the Bayesian paradigm from Chapter 2 and will in particular use MAP scores for CEGs as introduced in Section 2.1.4.

Let thus $\boldsymbol{M}_1, \boldsymbol{M}_2 \in \boldsymbol{C}$ be two competing models, represented by CEGs $\mathcal{C}_1$ and $\mathcal{C}_2$ with vectors of edge probabilities $\boldsymbol{\theta}_1$ and $\boldsymbol{\theta}_2$, respectively. Denote by $q(\mathcal{C}_i) = f_i(\boldsymbol{\theta}_i)$ the prior distribution we set for the CEG $\mathcal{C}_i$, $i = 1, 2$. In (2.18), we showed that when comparing two models, the ratio of their posterior probabilities is equal to their Bayes Factor multiplied by the ratio of their prior probabilities. So for our two CEGs, this is:

$$\frac{q_+(\mathcal{C}_1)}{q_+(\mathcal{C}_2)} = \frac{q(\mathcal{C}_1)}{q(\mathcal{C}_2)} \cdot \mathrm{BF}(\mathcal{C}_1, \mathcal{C}_2) \tag{6.1}$$

where $q_+(\mathcal{C}_i) = f_i(\boldsymbol{\theta}_i | \boldsymbol{y})$ denotes the respective posterior probability of the vector of edge labels given the data, $i = 1, 2$. In order to simplify computations, we will now take the logarithm on both sides of this equation just as we did when deriving the posterior log odds in (2.19). We thus obtain:

$$\log q_+(\mathcal{C}_1) - \log q_+(\mathcal{C}_2) = \log q(\mathcal{C}_1) - \log q(\mathcal{C}_2) + \log \mathrm{BF}(\mathcal{C}_1, \mathcal{C}_2) \tag{6.2}$$

The left hand side of (6.2) will henceforth be called the *log-posterior Bayes Factor (lpBF)* between the two models, denoted $\mathrm{lpBF}(\mathcal{C}_1, \mathcal{C}_2)$.

Explicitly, when we choose Dirichlet prior distributions with hyperparameters $\boldsymbol{\alpha}_1$ and $\boldsymbol{\alpha}_2$ over these vectors of edge probabilities, respectively, we can then write the log-posterior Bayes Factor between two CEG models in closed form:

$$
\begin{aligned}
\mathrm{lpBF}(\mathcal{C}_1, \mathcal{C}_2) =& \log q(\mathcal{C}_1) - \log q(\mathcal{C}_2) + \log f_1(\boldsymbol{y}) - \log f_2(\boldsymbol{y}) \\
=& \log q(\mathcal{C}_1) - \log q(\mathcal{C}_2) \\
&+ \sum_{i=1}^{\tilde{K}_1} \log \Gamma(\overline{\alpha}_{1i}) - \log \Gamma(\overline{\alpha}_{1i+}) - \sum_{j=1}^{K_1} \Big(\log \Gamma(\alpha_{1j}) - \log \Gamma(\alpha_{1j+}) \Big) \\
&- \sum_{i=1}^{\tilde{K}_2} \log \Gamma(\overline{\alpha}_{2i}) - \log \Gamma(\overline{\alpha}_{2i+}) - \sum_{j=1}^{K_1} \Big(\log \Gamma(\alpha_{2j}) - \log \Gamma(\alpha_{2j+}) \Big)
\end{aligned}
\tag{6.3}
$$

where $\tilde{K}_i$ denotes the total number of stages in the CEG $\mathcal{C}_i$ for $i = 1, 2$.

For 'neighbouring' models which are close in the sense that they have the same staged tree representation safe for one vertex identification, many terms in (6.3) cancel so that the relative score turns out to be particularly simple. We will do these calculations explicitly in Section 6.3.1.

Of course in the development above eliciting a prior probability $q(\mathcal{C})$ for each candidate CEG $\mathcal{C}$ representing a model $\boldsymbol{M}_{\mathcal{C}} \in \boldsymbol{C}$ is often a demanding task. This is because the size of the model space is immense even when there is only a single spanning event tree with a moderate number of situations: see the discussion in Section 6.4. For the purposes of explanatory data analysis, one popular choice is to assume all models are a priori equally likely. Formally, we would thus choose a *uniform prior* given by

$$
q(\mathcal{C}) = \frac{1}{\#\boldsymbol{C}} \qquad \text{for all } \mathcal{C} \in \boldsymbol{C} \tag{6.4}
$$

where $\#\boldsymbol{C}$ denotes the total number of models in the class $\boldsymbol{C}$. In this case, the log-posterior BF (6.3) between two CEGs reduces to the ratio of their marginal likelihoods. By (6.1), under this assumption the lpBF is thus *equal* to the logarithm of the BF score as introduced in (2.17). So $\mathrm{lpBF}(\mathcal{C}_1, \mathcal{C}_2) = \log \mathrm{BF}(\mathcal{C}_1, \mathcal{C}_2)$ for any two CEGs with the same prior. The highest scoring model in the CEG model space will then correspond to the CEG representing the MAP model: see Section 2.1.4.

An alternative method used for instance for BNs is to set a uniform prior not over the entire model class but over the class of statistically equivalent models. In the case of CEGs, this corresponds to setting

$$
q(\mathcal{C}) = \frac{1}{\#\mathcal{C}} \qquad \text{for all } \mathcal{C} \text{ representing a model } \boldsymbol{M}_{\mathcal{C}} \tag{6.5}
$$

where $\#\mathcal{C}$ denotes the total number of statistically equivalent CEGs representing the model $\boldsymbol{M}_{\mathcal{C}} \in \boldsymbol{C}$. Sadly in the case of CEGs this can be algorithmically much more complicated and still needs to be systematically developed, based on the recent characterisation of such statistical equivalence classes: see Section 4.2. However intuitive, these uniform priors are still not supported by some who argue that models can be distinguishable even though they embed the same conditional independence structures [62]. Thus in this chapter for simplicity we will use the popular uniform prior approach given in (6.4).

The MAP selection approach based on the lpBF score presented above provides a framework to conduct pairwise CEG model search quickly and easily. If it is possible for Bayesian model search to execute all possible comparisons over the whole class then the best scored model will be found eventually [22]. If this is not feasible then bespoke algorithms need to be designed so that it is possible to choose a good model, if not the best one. Comprehensive reviews about this issue as it relates to BF scores are given in [8, 61, 78]. For BF model selection over linear and classification tree models, expert systems and dynamic models, see for instance [15, 22] and [110], respectively.

In our case, we can compare two proposal CEGs $\mathcal{C}_1$ and $\mathcal{C}_2$ representing models in a class $\boldsymbol{C}$ in the same fashion by comparing their log-posterior Bayes Factors with those for a third CEG $\mathcal{C}_3$ in that same class:

$$\begin{aligned}\mathrm{lpBF}(\mathcal{C}_1, \mathcal{C}_2) &= \log f(\boldsymbol{\theta}_1|\boldsymbol{y}) - \log f(\boldsymbol{\theta}_3|\boldsymbol{y}) - \log f(\boldsymbol{\theta}_2|\boldsymbol{y}) + \log f(\boldsymbol{\theta}_3|\boldsymbol{y}) \\ &= \mathrm{lpBF}(\mathcal{C}_1, \mathcal{C}_3) - \mathrm{lpBF}(\mathcal{C}_2, \mathcal{C}_3).\end{aligned} \tag{6.6}$$

This property enables us to design an efficient model search algorithm using a heuristic strategy which restricts the model search space at each iteration to a local neighbourhood $\boldsymbol{N}(\mathcal{C}_3) \subseteq \boldsymbol{C}$, around the current best scored CEG $\mathcal{C}_3$. This neighbourhood will usually be determined as a collection of submodels. To select the best scored model in that neighbourhood it is necessary only to compare all models in $\boldsymbol{N}(\mathcal{C}_3)$ against $\mathcal{C}_3$. This often requires fewer calculations than computing the score associated with all pairs of models in $\boldsymbol{N}(\mathcal{C}_3)$ especially if all models in $\boldsymbol{N}(\mathcal{C}_3)$ are refinements of the model represented by $\mathcal{C}_3$. We give details of this in the next section.

6.3 CEG greedy and dynamic programming search

In this section we discuss the exhaustive and approximative model search algorithms developed for searching over a CEG model space. In both approaches, we distinguish between two possible cases usually found in real-world applications. The first case is when domain experts are able to fully construct the event tree that supports a CEG model [5]. The other situation arises when a family of event trees representing models in the space of interest is implicitly defined via a set of random variables $\boldsymbol{\mathcal{X}} = \{X_1, X_2, \ldots, X_m\}$ for some $m \geq 2$, each of which corresponds to a measurement made on each of the units that flow through the system we aim to model [23]. So this would be a family of $\boldsymbol{\mathcal{X}}$-compatible event trees as defined in Section 3.3 where the set $\boldsymbol{\mathcal{X}}$ is not a priori assumed to be ordered. We denote this family as

$\mathcal{T}_{\boldsymbol{\mathcal{X}}} = \{\mathcal{T} \mid \boldsymbol{M}_{\mathcal{T}} \in \boldsymbol{C}$ and $\mathcal{T}$ is $\boldsymbol{\mathcal{X}}$-compatible$\}$ and we will usually require the staging of these event trees to be stratified. Throughout this section, we will say that the event trees in $\mathcal{T}_{\boldsymbol{\mathcal{X}}}$ *span* the model search space $\boldsymbol{C}(\boldsymbol{\mathcal{X}}) \subseteq \boldsymbol{C}(\Omega)$ on the pre-specified set of atoms $\Omega = \times_{i=1}^{m} \mathbb{X}_i$ which is equal to the product-state space of the random variables $X_1, X_2, \ldots, X_m$.

Of course every tree $\mathcal{T} \in \mathcal{T}_{\boldsymbol{\mathcal{X}}}$ in the class of candidate explanations explicitly depicts the measurement variables in a certain order. To account for this we will introduce some extra notation. Thus let I always denote a permutation of the indices $(1, 2, \ldots, m) \mapsto (i_1, i_2, \ldots, i_m)$ which reorders the variables in $\boldsymbol{\mathcal{X}}$ into a vector $(X_{i_1}, X_{i_2}, \ldots, X_{i_m})$. We will henceforth denote that vector by $\boldsymbol{X}(I)$ and will denote an $\boldsymbol{X}(I)$-compatible event tree in the class $\mathcal{T}_{\boldsymbol{\mathcal{X}}}$ that represents this order as $\mathcal{T}(\boldsymbol{X}(I))$. Compare also the notation introduced in Section 3.3.

We can now present search algorithms for this class. Before we do so we briefly illustrate the new notation.

Example 6.1 (Liver and kidney disorders—two variables)**.** *Recall the example of patients diagnosed with liver and kidney disorders analysed through Sections 5.1.3 and 5.2.2. We now assume that a physician is interested in exploring how liver and kidney dysfunctions relate to each other. For this purpose, we define two binary random variables* X_{l} *and* X_{k} *that distinguish whether a patient has, respectively, a liver disorder or a kidney disorder. Then the set* $\boldsymbol{\mathcal{X}} = \{X_{\mathrm{l}}, X_{\mathrm{k}}\}$ *can be represented by two* $\boldsymbol{\mathcal{X}}$*-compatible event trees* $\mathcal{T}(X_{\mathrm{l}}, X_{\mathrm{k}}), \mathcal{T}(X_{\mathrm{k}}, X_{\mathrm{l}})$ *as depicted in Figure 6.1. In particular, Figure 6.1a shows the event tree* $\mathcal{T}(X_{\mathrm{l}}, X_{\mathrm{k}})$ *where the root situation is associated with liver disorder and all child situations correspond to kidney disorders and, conversely, Figure 6.1b depicts an event tree* $\mathcal{T}(X_{\mathrm{k}}, X_{\mathrm{l}})$ *exhibiting the inverse ordering. Here there are no domain clues to fix a variable order beforehand and the class of all possible explanations* $\mathcal{T}_{\boldsymbol{\mathcal{X}}}$ *is spanned by these two trees.*

In this setting it is reasonable to only search across models with merge only situations associated with the same random variable, so situations v_1 *and* v_2 *but not* v_0*, in both event trees. Under these assumptions any model within our CEG model search space* $\boldsymbol{C}(\boldsymbol{\mathcal{X}})$ *is represented by one of the four* $\boldsymbol{\mathcal{X}}$*-compatible stratified staged trees: one saturated and one where* v_1 *and* v_2 *are in the same stage for each variable order.*

6.3.1 Greedy SCEG search using AHC

CEG structure learning with a given event tree

In order to be able to search over a CEG model space which is spanned by a single given event tree, Freeman and Smith [33] implemented an *agglomerative hierarchical clustering (AHC)* algorithm using a Bayesian approach.

The general AHC method [49] organises data into a hierarchy of clusters. It starts from the singleton clusters defined by the data structure. Using some proximity score and usually adopting some greedy strategy the algorithm then

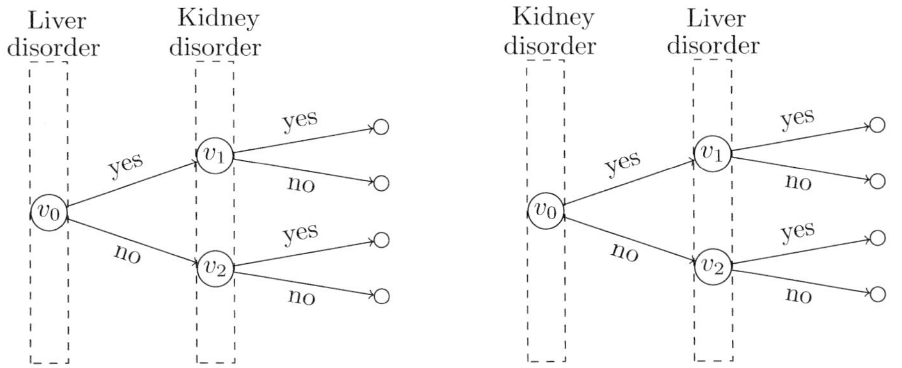

(a) Variable order $X_l \prec X_k$. **(b)** Variable order $X_k \prec X_l$.

FIGURE 6.1
Two $\boldsymbol{\mathcal{X}}$-compatible event trees explaining liver X_l and kidney X_k disorders. Here, $\boldsymbol{\mathcal{X}} = \{X_l, X_k\}$.

proceeds to merge sequentially the clusters at each iteration until obtaining a final cluster. Hereby the data is thus organised into a hierarchical sequence of nested partitions.

In the CEG context, the AHC algorithm initially starts from the saturated CEG $\mathcal{C}_0 = \mathcal{C}(\mathcal{T})$ whose underlying tree is precisely the given event tree $\mathcal{T}$. Using the Dirichlet characterisation of CEG models developed in the previous section, it then adopts a CEG greedy search strategy based on the model scores, so the log-posterior probability of each model. Explicitly, assume that the AHC algorithm has chosen the CEG $\mathcal{C}_i$ as the best local model at the end of an iteration i for some $i \in \{0, 2, \ldots, r-1\}$ where $r = \#\boldsymbol{C}$. Now define the local search neighbourhood at iteration $i+1$ as a family of models denoted $\boldsymbol{N}(\mathcal{C}_i) \subseteq \boldsymbol{C}$ which is constituted by all CEGs $\mathcal{C}(\mathcal{T}_{i+1})$ where $\mathcal{T}_{i+1} = \mathcal{T}$ has the same underlying event tree as the saturated model but with one stage less than $\mathcal{T}_i$. So if say u_1 and u_2 are distinct stages in T_i then $u_1 = u_2$ in $\mathcal{T}_{i+1}$. In this case we say that $\mathcal{C}(\mathcal{T}_{i+1})$ is *1-nested* in $\mathcal{C}(\mathcal{T}_i)$. For illustration, consider the two CEGs in Figure 6.2 depicting the liver and kidney disorder model from Example 6.1. Here, the CEG in Figure 6.2b is 1-nested into the saturated CEG in Figure 6.2a.

Setting a uniform prior (6.4) over the model space $\boldsymbol{C}$, the log-posterior Bayes Factor (6.3) between the model $\mathcal{C}_i$ and a candidate model $\mathcal{C}_{i+1} \in \boldsymbol{N}(\mathcal{C}_i)$

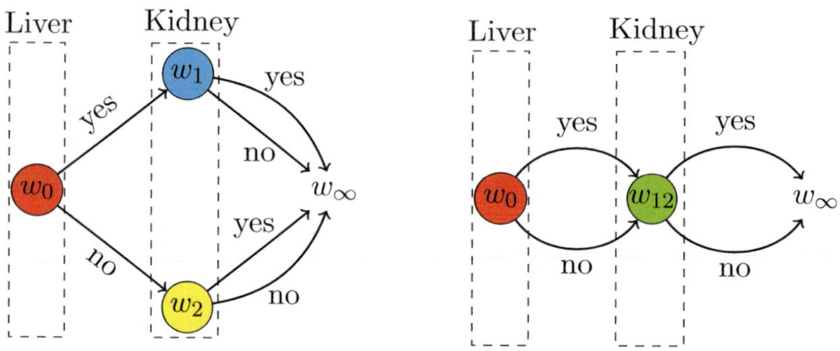

(a) The saturated CEG $\mathcal{C}_0$. **(b)** The 1-nested CEG $\mathcal{C}_1$.

FIGURE 6.2
The two possible CEGs that span the CEG model space associated with Example 6.1 when the variable ordering $X_{\rm l} \prec X_{\rm k}$ is assumed.

at iteration i can be simplified to be:

$$\begin{aligned} \text{lpBF}(\mathcal{C}_i, \mathcal{C}_{i+1}) = & \sum_{j=1}^{2} \log \Gamma(\overline{\alpha}_j) - \log \Gamma(\overline{\alpha}_{j+}) - \sum_{k=1}^{K_1} \Big(\log \Gamma(\alpha_{jk}) - \log \Gamma(\alpha_{jk+}) \Big) \\ & - \Big(\log \Gamma(\overline{\alpha}_1 + \overline{\alpha}_2) + \log \Gamma(\overline{\alpha}_{1+} + \overline{\alpha}_{2+}) \\ & - \sum_{k=1}^{K_1} \Big(\log \Gamma(\alpha_{1k} + \alpha_{2k}) - \log \Gamma(\alpha_{1k+} + \alpha_{2k+}) \Big) \Big) \end{aligned} \tag{6.7}$$

where u_1 and u_2 are the two stages of $\mathcal{C}_i$ that are merged to obtain $\mathcal{C}_{i+1}$, and the prior and posterior Dirichlet hyperparameters associated with stage u_j are denoted $\boldsymbol{\alpha}_{j1}$ and $\boldsymbol{\alpha}_{j+}$, respectively. The MAP CEG in the neighbourhood $\boldsymbol{N}(\mathcal{C}_i)$ is then simply the CEG $\mathcal{C}_{i+1}$ which maximises the score (6.7). This can simply be found using (6.6).

The AHC algorithm, given in pseudo-code in Algorithm 2, stops the search at the first iteration $i \leq r$ where the score $\text{lpBF}(\mathcal{C}_i, \mathcal{C}_{i-1})$ becomes negative. This has now been implemented in the R-package `ceg` [18].

SCEG structure learning without a given variable order

When the CEG model space we wish to search is a particular family of stratified CEGs compatible with $m \geq 2$ discrete random variables $\boldsymbol{\mathcal{X}} = \{X_1, X_2, \ldots, X_m\}$, the lack of a given variable order implies that this space is spanned by an enormous collection of $m!$ event trees $\boldsymbol{\mathcal{T}_{\mathcal{X}}}$. The computational complexity of searching this space will thus clearly be $O(m!)$. This search thus quickly becomes intractable even for moderately sized problems.

One possible way to circumvent this issue is to split the model search into two different steps: first find the best variable order and only afterwards look

---

Algorithm 2: AHC Algorithm

Input: A complete dataset $\boldsymbol{y}$, an event tree $\mathcal{T}$ and a hyperparameter $\overline{\alpha}$.
Output: The best scoring CEG found.

Initialise an array *stage* with the saturated stage structure of $\mathcal{C}_0 = \mathcal{C}(\mathcal{T})$.
Obtain the conditional frequency tables for each stage of $\mathcal{C}_0$ based on $\boldsymbol{y}$.
Calculate the hyperparameter $\boldsymbol{\alpha}$ for all stages of $\mathcal{C}_0$ using $\boldsymbol{y}$.
Initialise an array *score* with the log-posterior probabilities $Q(\mathcal{C}_0)$ of $\mathcal{C}_0$.
stop $\leftarrow$ FALSE
while *stop=FALSE* **do**
    for *every pair of stages* u_i *and* u_j *with the same number of outgoing edges* **do**
        Calculate the lpBF between the stage structure that merges the stages u_i and u_j into the same stage $u_{ij} = u_i \cup u_j$ keeping all other stages invariant and the stage structure *stage* using (6.7).
        if *there does not exist any pair* u_i *and* u_j **then**
            stop $\leftarrow$ TRUE
    Take the pair of stages $u_i^\star$ and $u_j^\star$ that provides the largest lpBF.
    if $\text{lpBF}[u_i^\star, u_j^\star] > 0$ **then**
        score $\leftarrow$ score + $\text{lpBF}[u_i^\star, u_j^\star]$
        Update *stage* gathering $u_i^\star$ and $u_j^\star$ into a single stage $u_{ij}^\star$.
    else
        stop $\leftarrow$ TRUE
return *stage, score*

---

for the best stage structure for the event tree compatible with that order. Because by the results developed in Section 3.3 every SCEG model is simply a submodel of a BN model, we can use any BN model search algorithm to discover a good variable order in this restricted class. For instance, in [5] the authors used an exhaustive search algorithm available in the R-package `deal` [11]. Of course, other algorithms based on exhaustive or heuristic approaches could be used as well. For an efficient dynamic programming algorithm to search the BN model space see [93].

The best scoring BN model will provide us with a variable ordering $\boldsymbol{X}(I) = (X_{i_1}, X_{i_2}, \cdots, X_{i_m})$ for some permutation I of the indices in the set of measurement variables $\boldsymbol{\mathcal{X}} = \{X_1, \ldots, X_m\}$. This ordering defines an $\boldsymbol{X}(I)$-compatible event tree $\mathcal{T}(\boldsymbol{X}(I))$ and hence also the set of SCEG models $\boldsymbol{C}(\boldsymbol{X}(I)) \subseteq \boldsymbol{C}(\boldsymbol{\mathcal{X}})$ specified over these trees. Now the AHC algorithm can be used to search over $\boldsymbol{C}(\boldsymbol{X}(I))$ for further asymmetric context-specific conditional statements that might be present in the data. As a result, the best scored SCEG will correspond to an embellishment of the best BN. This AHC

Algorithm 3: BN model search refined using the AHC Algorithm

Input: A complete data set $\boldsymbol{y}$ and a parameter $\overline{\alpha}$.
Output: The best scoring CEG found.

1. Find the best scoring BN.
2. Choose one of the Markov-equivalent variable orders I associated with the best scored BN.
3. Obtain the $\boldsymbol{X}(I)$-compatible event tree $\mathcal{T}(\boldsymbol{X}(I))$.
4. Find the best CEG using Algorithm 2 with inputs $\boldsymbol{y}, \mathcal{T}$ and $\bar{\alpha}$.

algorithm is given in pseudo-code in Algorithm 3 above and is now, just as the one for a fixed event tree as in the previous section, available in the `R`-package `ceg` [18].

Since one of the objectives in [5] was to compare the posterior probabilities between the embellished CEG and the best BN, they fixed a common value for the hyperparameter $\overline{\alpha}$ to initialise the CEG and BN model search algorithms. In general of course these values can be different. Sometimes this may be even desirable since the CEG model space is exponentially greater than the BN model space and so slightly greater values of this hyperparameter may enable us to obtain more stable results.

6.3.2 SCEG exhaustive search using DP

We now move from the AHC algorithms presented above which explore only local neighbourhoods of already found models to the more general approach of exhaustively searching a model space using *dynamic programming (DP)* techniques.

In this section the search is again restricted to the SCEG model space $\boldsymbol{C}(\boldsymbol{X}(I)) \subseteq \boldsymbol{C}(\Omega)$ defined by m discrete random variables $X_1, X_2, \ldots, X_m$ when a variable order $\boldsymbol{X}(I)$ as above is either given or has to be learned. We follow the Bayesian development presented in [23] which aims at finding the MAP SCEG model setting a uniform prior distribution (6.4). An alternative DP algorithm using non-Bayesian additive modular score was given in [92].

SCEG structure learning with a given variable order

In an $\boldsymbol{X}(I)$-compatible stratified CEG $\mathcal{C}$ with stage set $U = U_{\mathcal{C}}$, let $U_i \subseteq U$ denote the set of stages associated with the variable X_i for each $i = 1, \ldots, m$. Here, situations can be in the same stage only if they are associated with the same random variable. Therefore we can interpret an SCEG model represented by $\mathcal{C}$ as the intersection of m SCEG models, every single one of which is represented by a CEG which has non-trivial stages only along the ith level and is saturated otherwise. We will in the following denote by $Q(\mathcal{C}) = \log f(\boldsymbol{y})$ the log-marginal likelihood of the CEG $\mathcal{C}$ generating observational data $\boldsymbol{Y} = \boldsymbol{y}$ from complete random sampling. For SCEGs, this score may be written in

Algorithm 4: An exhaustive CEG model search given an event tree

Input: A complete data set $\boldsymbol{y}$, an ordered set of m variables $\boldsymbol{X}(I)$ and a hyperparameter $\overline{\alpha}$.

Output: The best scoring SCEG found.

Initialise an empty array U such that $\#U = m$.

Obtain the conditional frequency tables for each situation of $\mathcal{T}$ based on $\boldsymbol{y}$.

Calculate the hyperparameter $\boldsymbol{\alpha}$ for each situation of $\mathcal{T}$ using $\bar{\alpha}$ based on the conservative and uniform assumptions.

score $\leftarrow 0$

for *every variable in* $\boldsymbol{X}(I)$ **do**

  Calculate the local score Q_u of every possible subset of situations associated with X_i.

  Find the best scoring partition U_i given by Q_{U_i}.

  $U[i] \leftarrow U_i$

  score $\leftarrow$ score$+Q_{U_i}$

return *stage, score*

the decomposable form

$$Q(\mathcal{C}) = \sum_{i=1}^{m} Q_{U_i}(\mathcal{C}) \tag{6.8}$$

where $Q_{U_i}(\mathcal{C})$ denotes the log-marginal likelihood associated with variable X_i, so the score coming from the decomposition of an SCEG into its strata. Now (6.8) is again additively decomposable:

$$\begin{aligned} Q_{U_i}(\mathcal{C}) &= \sum_{u \in U_i} Q_u(\mathcal{C}) \\ &= \sum_{u \in U_i} \sum_{j=1}^{K_i} \log \frac{\Gamma(\alpha_{uj+})}{\Gamma(\alpha_{uj})} - \sum_{u \in U_i} \log \frac{\Gamma(\overline{\alpha}_{+u})}{\Gamma(\overline{\alpha}_u)} \end{aligned} \tag{6.9}$$

where $\boldsymbol{\alpha}_u$ denotes the Dirichlet hyperparameter associated to the stage u, $Q_u(\mathcal{C})$ now denotes the log-marginal likelihood of a stage $u \in U_i$ and K_i is the number of states of the variable X_i for all $i = 1, \ldots, m$. This formula is true because of the additive decomposition of the marginal likelihood of each CEG model given in (5.5).

We can therefore maximise the score of the SCEG $\mathcal{C} = \mathcal{C}(\boldsymbol{X}(I))$ level by level: maximising the score of each of its underlying variables, so each component of the vector $\boldsymbol{X}(I)$, independently. The pseudo-code for this procedure is given in Algorithm 4. Here, optimising the score associated with a variable X_i is achieved by computing the scores for every possible configuration of stages in U_i and then selecting the partition that provides us with the highest score, for all $i = 1, \ldots, m$.

Algorithm 5: Find the best scoring SCEG when no variable ordering is specified

---

Input: A complete data set $\boldsymbol{y}$ on a set of m finite discrete variables $\boldsymbol{\mathcal{X}}$ and a hyperparameter $\overline{\alpha}$.

Output: The best scoring SCEG found.

1 Discover the best sink variable for all 2^m non-empty subsets of $\boldsymbol{\mathcal{X}}$.

2 Find the best variable order $I^\star$.

3 Obtain the best SCEG using the Algorithm 4 with inputs $\boldsymbol{y}$, $\boldsymbol{X}(I^\star)$ and $\overline{\alpha}$.

SCEG structure learning without a given variable order

The additive modularity of the log-posterior probability of a SCEG guarantees that removing the last variable from an ordered set of variables does not change the actual best variable order for the remaining variables [23]. Explicitly, if $I^\star = (i_1^\star, \ldots, i_M^\star)$ denotes the highest scoring variable order for a CEG in the class $\boldsymbol{C}(\boldsymbol{\mathcal{X}})$ to represent a process described by the variable set $\boldsymbol{\mathcal{X}}$ then $I_{M-1}^\star = (i_1^\star, \ldots, i_{M-1}^\star)$, is the best variable order for a CEG model to express the subprocess corresponding to the variable set $\boldsymbol{\mathcal{X}} \setminus \{X_{i_M^\star}\}$ for $M \leq m$.

Example 6.2 (Liver and kidney disorders—three variables). *Consider again the setting in Example 6.1. Suppose that a physician wants to understand the interactions between the level of daily stress experienced by an individual and his risk of developing liver and kidney disorders. To model this process we define three random variables X_{l}, X_{k} and X_{s}. The first two variables X_{l} and X_{k} differentiate between patients who have a high, moderate or low risk of having* liver *disorder or* kidney *disorder, respectively. The last variable X_{s} distinguishes whether a patient lives under a high, moderate or low level of* stress. *So the set $\boldsymbol{\mathcal{X}} = \{X_{\mathrm{l}}, X_{\mathrm{k}}, X_{\mathrm{s}}\}$ gives rise to six $\boldsymbol{\mathcal{X}}$-compatible event trees representing the $3! = 6$ possible different orderings of these three variables.*

From (6.8) we deduce that

$$Q(\mathcal{C}) = \sum_{i=1}^{3} Q_{U_i}(\mathcal{C}) \qquad \text{for all CEGs } \mathcal{C} \text{ for which } \boldsymbol{M}_{\mathcal{C}} \in \boldsymbol{C}(\boldsymbol{\mathcal{X}}) \tag{6.10}$$

where U_1, U_2 and U_3 are the stage sets associated with variables X_{l}, X_{k} and X_{s}, respectively. Each score $Q_{U_i}(\mathcal{C})$ depends on the variable ordering. However, each highest scored stage structure U_i can be found independently from the others when a variable ordering is known as above, for some $i = 1, 2, 3$. In particular, if the variable ordering $\boldsymbol{X}(I^\star) = (X_{\mathrm{s}}, X_{\mathrm{k}}, X_{\mathrm{l}})$ provides the MAP SCEG $\mathcal{C}^\star$ representing the highest scored model in $\boldsymbol{C}(\boldsymbol{\mathcal{X}})$, then we have necessarily that the best variable ordering for the set $\boldsymbol{\mathcal{X}}_1^{(2)} = \{X_{\mathrm{s}}, X_{\mathrm{k}}\}$ with only two variables has to be $\boldsymbol{X}_1^{(2)}(I^\star) = (X_{\mathrm{s}}, X_{\mathrm{k}})$.

The observations above lead us to a recursive dynamic programming frame-

Algorithm 6: Find the best sink variables for every non-empty subset of $\boldsymbol{\mathcal{X}}$: step 1 of Algorithm 5

Input: A complete data set $\boldsymbol{y}$ on a set of m finite discrete variables $\boldsymbol{\mathcal{X}}$ and a hyperparameter $\overline{\alpha}$.

Output: A set-indexed array *sinks* that for each subset $\boldsymbol{\mathcal{X}}^k \subset \boldsymbol{\mathcal{X}}$ returns the sink variable for the highest scoring $\boldsymbol{\mathcal{X}}^k$-compatible CEG.

for k *in* $1 \to n$ **do**
    for $\boldsymbol{\mathcal{X}}^k \subset X$ *such that* $|\boldsymbol{\mathcal{X}}^k| = k$ **do**
        scores$[\boldsymbol{\mathcal{X}}^k] \leftarrow 0$
        sinks$[\boldsymbol{\mathcal{X}}^k] \leftarrow -1$
        for $X_i \in \boldsymbol{\mathcal{X}}^k$ **do**
            $\boldsymbol{\mathcal{X}}^{k-1} \leftarrow \boldsymbol{\mathcal{X}}^k \setminus \{X_i\}$
            score$L \leftarrow$ BLS$(X_i, \boldsymbol{\mathcal{X}}^{k-1})$+scores$[\boldsymbol{\mathcal{X}}^{k-1}]$
            if *sinks*$[\boldsymbol{\mathcal{X}}^k] = -1$ *or scoreL* $>$ *scores*$[\boldsymbol{\mathcal{X}}^k]$ **then**
                scores$[\boldsymbol{\mathcal{X}}^k] \leftarrow$score$L$
                sinks$[\boldsymbol{\mathcal{X}}^k] \leftarrow X_i$

return *sinks*

work where the problem of finding the best variable order for $M-1$ variables constitutes a subproblem of discovering the best variable order for N variables. Therefore, for every subset $\boldsymbol{\mathcal{X}}^k = \{X_{i_1}, \ldots, X_{i_k}\} \subseteq \boldsymbol{\mathcal{X}}$ of the given problem variables, $k = 1, \ldots, M-1$, we have to find the best sink variable $X_i \in \boldsymbol{\mathcal{X}}^k$ given that we have already found the best variable order for every subset $\boldsymbol{\mathcal{X}}^{k-1} \subseteq \boldsymbol{\mathcal{X}}^k$. Embedding this recursive structure into a DP algorithm enables us to search efficiently the entire SCEG model space. The general algorithm for learning SCEGs is given again in pseudo-code in Algorithm 5 and has been implemented in the `R`-package `ceg` [18]. We dedicate the remainder of this section to an explanation and illustration of the three steps that constitute this algorithm.

Step 1: Discover the best sink variable Algorithm 6 constitutes the most computationally intense step of the general DP algorithm for CEG model search. It begins initialising two 2^m-size arrays called *scores* and *sinks*, each element of which corresponds to a subset of $\boldsymbol{\mathcal{X}}$. It then proceeds to determine the best sink variable of each non-empty subset of $\boldsymbol{\mathcal{X}}$ by examining them in order of increasing size, starting with singleton subsets.

For every variable X_i in a set $\boldsymbol{\mathcal{X}}^{k+1}$ it is necessary to calculate the local score of the best staged tree spanned by the set of variables $\boldsymbol{\mathcal{X}}^k \cup \{X_i\}$ such that X_i is the sink variable. To do this, the algorithm first requires a local auxiliary variable score and a function BLS$(X_i, \boldsymbol{\mathcal{X}}^k)$. This *best local score* associated with $\boldsymbol{\mathcal{X}}^k$ has already been computed and stored in *score* since the algorithm looks at subsets ordered by increasing size. So the function $\text{BLS}(X_i, \boldsymbol{\mathcal{X}}^k)$ only needs to calculate the score Q_{U_i} of the best stage partition U_i associated with the sink variable X_i. Observe that this does not require the best variable ordering of $\boldsymbol{\mathcal{X}}^k$.

Example 6.3 (Example 6.2 continued). *Suppose that the physician now asks his analyst collaborator to model the stress-disease problem. The analyst decides to present the MAP SCEG* $C^\star$ *to the physician. For this purpose, he uses the DP-algorithm given in Algorithm 5. In the first step, he needs to find a triad* $(\boldsymbol{\mathcal{X}}^k, X_\star, \vartheta_k)$ *for every set* $\boldsymbol{\mathcal{X}}^k \subseteq \boldsymbol{\mathcal{X}}$*, where* $X_\star$ *denotes the best sink variable for* $\boldsymbol{\mathcal{X}}^k$ *and* ϑ_k *denotes the best local score associated with* $\boldsymbol{\mathcal{X}}^k$*. The algorithm starts from singleton subsets, so from* $(\{X_\text{l}\}, X_\text{l}, \vartheta_1)$*,* $(\{X_\text{k}\}, X_\text{k}, \vartheta_2)$ *and* $(\{X_\text{s}\}, X_\text{s}, \vartheta_3)$*.*

Next the analyst examines the sets of size two. For instance, take the set $\boldsymbol{\mathcal{X}}_2^{(2)} = \{X_\text{l}, X_\text{k}\}$*. To find the best sink variable for this subset it is necessary to compare the highest scored staged tree associated with the variable ordering* $\boldsymbol{X}_2^{(2)}(I_1) = (X_\text{l}, X_\text{k})$ *against the best tree resulting from the alternative variable ordering* $\boldsymbol{X}_2^{(2)}(I_2) = (X_\text{k}, X_\text{l})$*: compare Figure 6.1. Note that for the variable ordering* $\boldsymbol{X}_2^{(2)}(I_1)$ *we have to compute only the best score* ϑ_a *associated with* X_k *since the score of* X_l *has already been computed previously and it is equal to* ϑ_1*. In analogy, for the variable ordering* $\boldsymbol{X}_2^{(2)}(I_2)$ *we have to find only the best score* ϑ_b *for* X_l*. Now assume that* $\vartheta_2 + \vartheta_a < \vartheta_1 + \vartheta_b$*. So with regard to the set* $\boldsymbol{\mathcal{X}}_2^{(2)}$ *the best sink variable is* X_l *and its highest score is* $\vartheta_4 = \vartheta_2 + \vartheta_a$*. Following this approach also for the other two subsets of size two, the algorithm obtains the following triads:* $(\{X_\text{l}, X_\text{k}\}, X_\text{l}, \vartheta_4)$*,* $(\{X_\text{l}, X_\text{s}\}, X_\text{s}, \vartheta_5)$ *and* $(\{X_\text{k}, X_\text{s}\}, X_\text{k}, \vartheta_6)$*.*

To finalise Step 1, the analyst now uses the algorithm to search for the best sink variable in the set $\boldsymbol{\mathcal{X}}$*. Of course, there are three possible candidates:* X_l*,* X_k *and* X_s*. For example, take the variable* X_l*. In this case, we have to find the score of the MAP staged tree which has* X_l *as its sink variable. This corresponds only to computing the score* ϑ_c *of* X_l *and then adding it to the score* ϑ_6 *that was previously calculated since the best variable ordering of a staged tree does not change if the sink variable is eliminated. Repeating the same procedure for the other two candidate variables we can obtain the score* ϑ_d *for* X_k *and* ϑ_e *for* X_s*. Now assume that* $\vartheta_6 + \vartheta_c > \vartheta_5 + \vartheta_d > \vartheta_e + \vartheta_4$*. It then follows that the algorithm stores the triad* $(\{X_\text{l}, X_\text{k}, X_\text{s}\}, X_\text{l}, \vartheta_7)$*, where* $\vartheta_7 = \vartheta_6 + \vartheta_c$*.*

Step 2: Find a best ordering of the best sinks Now the pseudo-code in Algorithm 7 shows how to find the best ordering of the best sink variables starting with the complete set $\boldsymbol{\mathcal{X}}$. For this purpose at iteration k we have to determine iteratively the best sink variables for the auxiliary set of variables $\boldsymbol{\mathcal{X}}_\text{left}$ described below, for $k = m, \ldots, 1$.

The set $\boldsymbol{\mathcal{X}}_\text{left}$ is initialised with $\boldsymbol{\mathcal{X}}$. At each iteration k the algorithm first recovers the best sink variable X_{i_k} of $\boldsymbol{\mathcal{X}}_\text{left}$ from the indexed array *sinks* given

Algorithm 7: Find the best variable ordering: step 2 of Algorithm 5

Input: The set indexed array *sinks*.

Output: An integer-indexed array of the variable ordering for the highest scoring CEG.

$\boldsymbol{\mathcal{X}}_{\text{left}} = \boldsymbol{\mathcal{X}}$

for *i from m to* 1 **do**

  order[i]$\leftarrow$sinks[$\boldsymbol{\mathcal{X}}_{\text{left}}$]

  $\boldsymbol{\mathcal{X}}_{\text{left}} \leftarrow \boldsymbol{\mathcal{X}}_{\text{left}} \setminus \{\text{order}[i]\}$

return *order*

---

by Algorithm 6. Next it removes X_{i_k} from $\boldsymbol{\mathcal{X}}_{\text{left}}$ and sets $\boldsymbol{\mathcal{X}}_{\text{left}} = \boldsymbol{\mathcal{X}}_{\text{left}} \setminus \{X_{i_k}\}$. It then begins the iteration $k-1$. The variable X_{i_k} is stored as the kth element of an m-dimensional integer indexed array *order* of variables.

By carrying out these algorithmic interactions in decreasing order, it then follows that in the end the array *order* contains the variable ordering for the highest scoring SCEG. It is also straightforward to see that the root variable X_{i_1} corresponds to the variable order[1] whilst the last variable X_{i_m} is stored in order[m]. The computational complexity of this step is linear in m.

Example 6.4 (Example 6.3 continued)**.** *In the second step the analyst now employs the DP algorithm to find the best variable ordering. Using the triad calculated in the previous step, this task is very simple. Starting with the full set* $\boldsymbol{\mathcal{X}}$*, we first derive that the best sink variable is* X_{k}*. Then the algorithm identifies the best sink variable for the remaining set* $\boldsymbol{\mathcal{X}}_{left} = \boldsymbol{\mathcal{X}} \setminus \{X_{\text{l}}\}$*. The result of this is* X_{k} *since* $\boldsymbol{\mathcal{X}}_{left} = \{X_{\text{l}}, X_{\text{k}}\}$*. Finally, the updated set* $\boldsymbol{\mathcal{X}}_{left} = \boldsymbol{\mathcal{X}}_{left} \setminus \{X_{\text{l}}\} = \{X_{\text{s}}\}$ *is a singleton set and so the algorithm terminates here. The best variable ordering is then given by* $\boldsymbol{X}(I^{\star}) = (X_{\text{s}}, X_{\text{k}}, X_{\text{l}})$*.*

Step 3: Recover the highest scoring SCEG This final step is the simplest of the three steps in Algorithm 5. In fact, having already inferred the best variable order, we now only need to apply Algorithm 4 to recover the highest scoring SCEG for our input data. After this step, the DP algorithm terminates.

The DP algorithm for CEG model search we have presented above closely resembles the DP algorithm for BN learning [93]. The main difference is that in the DP algorithm for BN model selection there is a pre-processing step where all local scores are pre-computed. Therefore the MAP BN can be recovered quite directly and at little extra cost since we do not need to run an algorithm given the best BN variable ordering to find the best parent configuration for each variable. This is because the parent set associated with each variable is actually stored in memory by the algorithm. Instead of recalculating this quantity, the DP algorithm for CEG learning calculates the local scores as required and caches them. Despite the additional computational time required by the third step to recover the best CEG, this three-step approach adopted for CEG model search is justified because of its much reduced memory cost and also its computational simplicity.

The SCEG model space is far larger than the BN model space and so there are many more partitions in it whose scores should have to be computed and stored. In the BN framework a local score for any variable is calculated based on an unordered set of its parents. This implies that given a variable X_k and a subset of variables $\boldsymbol{\mathcal{X}}^k$ storing the best set of parents for X_k in $\boldsymbol{\mathcal{X}}^k$ together with the best local score is computationally cheap. The same observation does not hold for CEGs because whilst the variable ordering of $\boldsymbol{\mathcal{X}}^k$ does not change the score Q_{U_k} associated with the sink variable X_k it does alter its stage structure U_k. In fact, different variable orderings permute the leaf nodes of the event tree spanned by $\boldsymbol{\mathcal{X}}^k \cup \{X_k\}$ where X_k is the final variable in the ordering. Therefore, a fast recovery of the MAP SCEG would require to store the best stage configuration for every pair $(X_k, \boldsymbol{X}^k)$, where $\boldsymbol{X}^k$ is a possible ordered sequence of $\boldsymbol{\mathcal{X}}^k$. This would add a complexity of factorial order in the algorithm which often exceeds the cost of running Algorithm 4 to recover the MAP SCEG.

In the next chapter we will illustrate the practical outworking of the model selection algorithms described above. We will compare these with two more recent methodologies described below. These appear to perform even better. For completeness we thus briefly describe these below.

6.4 Technical advances for SCEG model selection

In this last part of this chapter we will briefly discuss two advancements of the methods described above. The first introduces how partial information about the ordering of variables, or more generally of situations, can be embedded in a model search algorithm. The second is a recent methodology based on *non-local priors (NLPs)* which appears to outperform standard MAP search of the SCEG model space in two ways. First the method tends to select simpler models more quickly, usefully enforcing more parsimony. Second the method appears to be extremely robust to the choice of the effective sample size parameter discussed in Sections 5.1.2 and 6.1.

6.4.1 DP and AHC using a block ordering

When learning a BN we learn a restricted set of partitions of measurement variables. This prevents us from exploring the context-specific conditional independences and possible asymmetries in the development of a process. In contrast, the CEG model space is structurally more flexible and this advantage comes at a computational cost for CEG model selection.

To see this consider a CEG model space $\boldsymbol{C}(\boldsymbol{\mathcal{X}})$ spanned by a set of $m \geq 2$, discrete random variables $\boldsymbol{\mathcal{X}} = \{X_1, X_2, \ldots, X_m\}$, where each random variable X_i has a finite number of $K_i = \#\mathbb{X}_i$ states. Let $\kappa_i = \prod_{j=1}^{k} K_{i_j}$ be the number of situations associated with variable X_i in an event tree $\mathcal{T}(\boldsymbol{X}(I))$. The total number of partitions of these situations corresponding to variable X_i is then given by the κ_i^{th} *Bell number* B_{κ_i} [101]. Now each partition constitutes a different stage structure U_i. This in turn corresponds to a distinct SCEG model regarding to a variable X_i for each $i = 1, \ldots, m$. Considering the $m!$ variable orderings it then follows that the size of such a SCEG model space can be expressed as

$$\#\{\mathcal{C} \text{ SCEG} \mid \boldsymbol{M}_{\mathcal{C}} \in \boldsymbol{C}(\boldsymbol{\mathcal{X}})\} = \sum_{I \in S_m} \prod_{i \in I} B_{\kappa_i}, \tag{6.11}$$

where S_m is the set of all possible permutations I of the indices $1, 2, \ldots, m$.

This implies that the complexity of this space grows exponentially in terms of Bell numbers. It depends on not only the number of variables but also the number of states that each of these variables has. Therefore, searching over the SCEG model space is enormously more computationally challenging than searching over its corresponding BN model space. For example, consider a process defined by a set of four binary random variables whose ordering is known. Learning a BN model requires us to calculate only $15 = \sum_{i=1}^{4} 2^{i-1}$ local scores whilst learning a SCEG model entails the computation of $4,158 = \sum_{i=1}^{4}(B_{2^{i-1}} - 1)$ local scores.

Therefore, the DP search method presented in Section 6.3 quickly becomes infeasible as the number of random variables in the set $\boldsymbol{\mathcal{X}}$ increases to an even moderate size. In this case, heuristic search strategies such as *agglomerative clustering* are needed to scale down the size of the SCEG model space to search over [23, 92]. A promising fast approximation is to embed this heuristic within the DP algorithm as done by Silander et al. [92]. These authors were able to search over model space defined by up to 18 random variables in less than 10 minutes. They also showed empirically that the AHC approach performed better than K-mean clustering methods (see e.g. [1, 55, 111]) when they are used in conjunction with the DP model search. However, the AHC algorithm is still much slower.

We will now show how to implement some of these advancements and how to combine them with domain knowledge.

To use agglomerative clustering, it is necessary only to rewrite the best local score $\text{BLS}(X_i, \boldsymbol{\mathcal{X}}^k)$ used in Algorithm 6. Instead of looking at the scores of all possible stage structures, this function will now find the best stage partition U_i associated with the variable X_i in a set $\boldsymbol{\mathcal{X}}^k$ using the adopted heuristic algorithm in Algorithm 8 for $i = 1, \ldots, m$.

During the modelling process, the identification of a feasible partial ordering for the variables in $\boldsymbol{\mathcal{X}}$ based on domain information may enable modellers to considerably reduce the computational complexities in full search methods.

Algorithm 8: Find the best sink variables for every non-empty subset of $\boldsymbol{\mathcal{X}}$ consistent with a block ordering $\boldsymbol{\mathcal{B}}$

Input: A complete dataset $\boldsymbol{y}$ on a set of m finite discrete variables $\boldsymbol{\mathcal{X}}$, a block ordering $\boldsymbol{\mathcal{B}} = (\boldsymbol{\mathcal{B}}_1, \ldots, \boldsymbol{\mathcal{B}}_l)$ of $\boldsymbol{\mathcal{X}}$ and a hyperparameter $\bar{\alpha}$.

Output: A set-indexed array *sinks* that for each subset $\boldsymbol{\mathcal{X}}^\star \subset \boldsymbol{\mathcal{X}}$ consistent with the block ordering returns the sink variable for the highest scoring SCEG spanned by $\boldsymbol{\mathcal{X}}^l$.

for *i from* 1 *to* l **do**
    for k *from* 1 *to* $\#\boldsymbol{\mathcal{B}}_i$ **do**
        for $\boldsymbol{\mathcal{B}}_i^k \subset \boldsymbol{\mathcal{B}}_i$ *such that* $\#\boldsymbol{\mathcal{B}}_i^k = k$ **do**
            $\boldsymbol{\mathcal{X}}^\star = \bigcup_{j=0}^{i-1} \boldsymbol{\mathcal{B}}_j \cup \boldsymbol{\mathcal{B}}_i^k$ where $\boldsymbol{\mathcal{B}}_0 = \emptyset$
            scores$[\boldsymbol{\mathcal{X}}^\star] \leftarrow 0$
            sinks$[\boldsymbol{\mathcal{X}}^\star] \leftarrow -1$
            for $X_i \in \boldsymbol{\mathcal{B}}_b^k$ **do**
                $\boldsymbol{\mathcal{X}}^{\star-1} \leftarrow \boldsymbol{\mathcal{X}}^\star \setminus \{X_i\}$
                score$L \leftarrow$ BLS$(X_i, \boldsymbol{\mathcal{X}}^{\star-1})$+scores$[\boldsymbol{\mathcal{X}}^{\star-1}]$
                if *sinks*$[\boldsymbol{\mathcal{X}}^\star] = -1$ *or scoreL* $>$*scores*$[\boldsymbol{\mathcal{X}}^\star]$ **then**
                    scores$[\boldsymbol{\mathcal{X}}^\star] \leftarrow$ score$L$
                    sinks$[\boldsymbol{\mathcal{X}}^\star] \leftarrow X_i$

return *sinks*

Particularly, the definition of a block ordering such that the blocks are well-ordered and constitute a partition of $\boldsymbol{\mathcal{X}}$ greatly reduces the space of allowed partitions that the search needs to be carried out on.

Let thus $\boldsymbol{\mathcal{B}} = (\boldsymbol{\mathcal{B}}_1, \ldots, \boldsymbol{\mathcal{B}}_l)$ denote a block ordering of the variables in $\boldsymbol{\mathcal{X}}$ where each $\boldsymbol{\mathcal{B}}_i = \{X_{i_1}, \ldots, X_{i_{l_i}}\} \subseteq \boldsymbol{\mathcal{X}}$ and every two blocks $\boldsymbol{\mathcal{B}}_i$ and $\boldsymbol{\mathcal{B}}_j$ are disjoint, $i \neq j$. The highest scoring SCEG can then be found among those obtained by permuting the variables within each block $\boldsymbol{\mathcal{B}}_i$ for $i = 1, \ldots, l$. Algorithm 8 implements this procedure by adding a loop in Algorithm 6 to control for the blocks. Note that the function BLS and the other steps of the algorithm do not change.

Parallel computation is a good option to speed up the exhaustive model searches. The key observation here is that the local scores $Q_{U_{ik}}$ associated with a variable X_{i_k} at level k in the underlying event tree can be independently computed from the local scores of variables at other levels. The speed-up gain of this approach can be substantial especially for the last levels of large event trees. When a variable ordering is known, the loop over the sequence of variables $\boldsymbol{X}(I)$ in line 5 of Algorithm 4 can be directly parallelised. In case of a full search without a variable ordering, parallel programming can be easily implemented over the intra-level loop to find the best sink variables. This corresponds to parallelising the computation of the inner loop over the set of variables $\boldsymbol{\mathcal{X}}_k$ in line 5 of Algorithm 6. When a bock ordering is given, parallel computation can also be introduced over the blocks in line 2 of Algorithm 8 and inside the blocks in line 7 of Algorithm 8.

The exhaustive SCEG model search algorithm can be easily customised to a general CEG model search with a given event tree that does not need to be an $\boldsymbol{\mathcal{X}}$-compatible event tree. For this purpose, it will be necessary to have a well-ordered partition $\boldsymbol{S} = \{\boldsymbol{S}_1, \ldots, \boldsymbol{S}_r\}$ over the set of non-leaf vertices of the tree. Algorithm 4 can then be directly used if $\boldsymbol{X}(I)$ is replaced by $\boldsymbol{S}$. Note that when the event tree is $\boldsymbol{\mathcal{X}}$-compatible the partition $\boldsymbol{S}$ is implicitly defined by the set of situations associated with the same variable. For instance, it is straightforward to see that the event tree in Figure 5.1a for the original liver-kidney-disorder model is not $\boldsymbol{\mathcal{X}}$-compatible. However, the partition $\boldsymbol{S}$ can be naturally defined according to the different geometric shapes of each situation in that tree:

$$\{\boldsymbol{S}_1 = \{s_0\}, \boldsymbol{S}_1 = \{s_1, s_2, s_3\}, \boldsymbol{S}_2 = \{s_4, s_5, s_7\}, \boldsymbol{S}_3 = \{s_6, s_8\}, \boldsymbol{S}_4 = \{s_9, s_{10}\}\}.$$

An obvious improvement of the AHC algorithm is to elicit a partition $\boldsymbol{S}$ as above and to only merge stages whose situations are associated with the same set in that partition as in line 7 of Algorithm 2. This is a natural restriction in many real-world problems. For example, in a SCEG model space we would often prefer not to gather situations corresponding to different random variables—although of course this is possible if the situations have the same number of outgoing edges.

Another possible embellishment to the AHC algorithm is to enlarge its search neighbourhood. This enables us to cover larger parts of the CEG model space and so increases the chance of finding a highest scoring CEG. However, this increases the computational cost substantially. For instance, adopting a partition $\boldsymbol{S}$, the worst case complexity of the search associated with each set $\boldsymbol{S}_i \in \boldsymbol{S}$ is $O(\#\boldsymbol{S}_i^{t+1})$ where t denotes the number of elements in the chosen neighbourhood. Of course, parallel computing can help us to keep the computational time under control as t increases because the AHC algorithm can again also be parallelised over the sets $\boldsymbol{S}_i$ for all $i = 1, \ldots, r$.

6.4.2 A pairwise moment non-local prior

To select between nested models $\boldsymbol{M}_1 \subseteq \boldsymbol{M}_0$ whose CEG representations $\mathcal{C}_0$ and $\mathcal{C}_1$ are nested as in Section 6.3.1, most applied Bayes Factor methods use prior probability distributions that define the null model's parameter space as contained within the alternative model's parameter space. Here this would correspond to $\mathcal{C}_1$ being 1-nested in $\mathcal{C}_0$. These prior distributions are called *local priors (LPs)* and are often based on conjugate priors as in Section 5.1.1. Recent studies on LPs [27, 28, 58] have shown that under a true alternative model $\boldsymbol{M}_1$ the collection of evidence grows exponentially whilst under a true null model $\boldsymbol{M}_0$ it increases only polynomially. This imbalance in the learning rate implies that BF techniques based on LPs are prone to choose more complex models than the true one.

To outsmart this issue, BF selection methods based on *non-local priors (NLPs)* have been successfully developed for linear models [58, 59] and graphs of Gaussian variables [2, 19, 20]. A NLP of a larger null model vanishes when the values of its parameters get close to the parameter space corresponding to a simpler alternative model, so if $\lim_{\boldsymbol{\theta}\to\boldsymbol{\theta}_0} f_1(\boldsymbol{\theta}) = 0$ for any $\boldsymbol{\theta}_0$ in the parameter space of $\boldsymbol{M}_0$ and f_1 being the prior distribution for $\boldsymbol{M}_1$.

One of the great advantages of a NLP is that it incorporates a notion of separation between two nested models directly in the prior distribution. In doing this, it scales up under the true null model without modifying the learning rate under the true alternative model. This characteristic robustifies inference by embodying NLPs with the prior belief that the data generation mechanism is defined by a parsimonious model. This then facilitates the recovery of high-dimensional conditional dependence structures when these really do drive the process: compare Definition 6.5 and Theorem 4 below.

We have shown that LPs are also prone to select more complex models when BF is used for CEG model selection [17]. However, this is not the only expected inconvenience with LPs in the CEG setting, particularly when some agglomerative model search algorithm is used. In such greedy searches, stages that are only visited rarely tend to be merged into stages that are more likely to be visited regardless of the data generating structures that define the conditional probability distributions of these stages. NLPs look like a promising alternative method for CEG model selection using a greedy search engine, such as the AHC algorithm, for two main reasons. First, they enable us to enforce parsimony and stability over a model search with respect to the setting of the hyperparameter. Second, by taking into consideration the distance between the probability distributions associated with two different stages they also discourage an agglomerative search method from merging two stages spuriously simply because of their probability density function.

We can now briefly outline some properties of the *pairwise moment* NLPs which constitute a new class of NLPs specifically customised for discrete processes developing over trees. For an extensive discussion on pm-NLPs and other two families of NLPs for CEGs see [17].

Definition 6.5 (Pairwise moment non-local priors for CEGs)**.** *Consider the CEG graph $\mathcal{C}_i$ and its 1-nested CEG graph $\mathcal{C}_{i+1}$ where the stages u_1 and u_2 are merged into $u_{12} = u_1 \cup u_2$. Denote by $\boldsymbol{\theta}_i$ the floret parameter vector of the stage u_i for $i = 1, 2, 12$. Then to test the model represented by $\mathcal{C}_i$ against the model represented by $\mathcal{C}_{i+1}$, we define the* pairwise moment non-local prior (pm-NLPs) *for $\mathcal{C}_{i+1}$ to be equal to a Dirichlet local prior $q_{\text{LP}}(\boldsymbol{\theta}_{12})$ and the priors for the finer CEG $\mathcal{C}_i$ to be given by*

$$q_{\text{NLP}}(\mathcal{C}_i) = \frac{1}{c_{12}} d(\boldsymbol{\theta}_1, \boldsymbol{\theta}_2)^{2\rho} q_{\text{LP}}(\boldsymbol{\theta}_{12}) \tag{6.12}$$

where $c_{12} = \mathbb{E}_{\boldsymbol{\theta}_1,\boldsymbol{\theta}_2}[d(\boldsymbol{\theta}_1,\boldsymbol{\theta}_2)^{2\rho}]$ *is the normalisation constant,* d *denotes a chosen distance and* $\rho \in \mathbb{N}$.

Using pm-NLPs for CEG model selection ensures that the parameters of the complex model have pm-NLP distributions whilst the parameters of the simple model have Dirichlet LP distributions. So the prior to be plugged into the complex model relies on the simple one. This inconsistency therefore requires a prior elicitation of the event tree. Note that when using the AHC algorithm this is not a critical requirement because an event tree is already given as an input.

Example 6.6 (Example 6.1 revisited)**.** *To illustrate how we might set NLPs for CEGs, return to the two-variable model we designed for liver and kidney dysfunctions. Assume that the CEG model space* $\boldsymbol{\mathcal{X}}$ *is spanned by a single event tree compatible with the variable order* $X_{\mathrm{l}} \prec X_{\mathrm{k}}$. *This was depicted in Figure 6.1a. Then in* $\boldsymbol{C}(\boldsymbol{\mathcal{X}})$ *there are only two possible SCEGs: the saturated one* $\mathcal{C}_0$ *and the one representing independence, here denoted* $\mathcal{C}_1$. *We have depicted these graphs in Figure 6.2. We can see in Figure 6.2a that the saturated CEG* $\mathcal{C}_0$ *has two different stages associated with the random variable* X_{k}, *namely* $u_1 = \{w_1\} = \{v_1\}$ *and* $u_2 = \{w_2\} = \{v_2\}$. *In contrast, as illustrated in Figure 6.2b, the simpler* 1*-nested CEG* $\mathcal{C}_1$ *has only one stage* $u_{12} = \{w_{12}\} = \{v_1, v_2\}$ *where the previous stages* u_1 *and* u_2 *of* $\mathcal{C}_0$ *have been gathered into a new single stage* u_{12}.

To select between these two models, we need only to determine whether the stages u_1 *and* u_2 *should be merged into a single stage* u_{12} *or whether they should be kept apart. This corresponds to testing the null hypotheses* $H_0 : \boldsymbol{\theta}_1 = \boldsymbol{\theta}_2$ *against the alternative* $H_1 : \boldsymbol{\theta}_1 \neq \boldsymbol{\theta}_2$ *where as above* $\boldsymbol{\theta}_i$ *denotes the parameter vector of the stage* u_i *for* $i = 1, 2$.

Now suppose that we would like to set Dirichlet pm-NLPs for these two models. Since $\mathcal{C}_1$ *is the simpler graph that can be obtained, its stage* u_{12} *can under the assumptions on stratified CEGs not be merged with any other stage. As a consequence, its corresponding NLP is identical to its Dirichlet LP,* $q_{\mathrm{NLP}}(\boldsymbol{\theta}_{12}) = q_{\mathrm{LP}}(\boldsymbol{\theta}_{12})$*: see Figure 6.3. To construct the NLPs for stages* u_1 *and* u_2 *of the saturated CEG* $\mathcal{C}_0$ *we need to combine the distance* $d(\boldsymbol{\theta}_1, \boldsymbol{\theta}_2)$ *between these two stages and their standard Dirichlet LPs* $q_{\mathrm{LP}}(\boldsymbol{\theta}_1)$ *and* $q_{\mathrm{LP}}(\boldsymbol{\theta}_2)$ *as in Definition 6.5:*

$$q_{\mathrm{NLP}}(\boldsymbol{\theta}_1, \boldsymbol{\theta}_2) = \frac{1}{c_{12}} d(\boldsymbol{\theta}_1, \boldsymbol{\theta}_2)^{2\rho} q_{\mathrm{LP}}(\boldsymbol{\theta}_1) q_{\mathrm{LP}}(\boldsymbol{\theta}_2), \tag{6.13}$$

where the proportionality constant $c_{12} = \mathbb{E}_{\boldsymbol{\theta}_1,\boldsymbol{\theta}_2}[d(\boldsymbol{\theta}_1,\boldsymbol{\theta}_2)^{2\rho}]$ *is computed with respect to the Dirichlet LPs and* $\rho = 1, 2, \ldots,$ *is a hyperparameter that controls the separation measure between models represented by hypotheses* H_0 *and* H_1 *given an adopted distance* d.

Figure 6.4 depicts the joint local and non-local prior probability distributions for the parameters $\theta_{1,1}$ *and* $\theta_{2,1}$ *in* $\mathcal{C}_0$ *which corresponds to the chance of a positive diagnosis of a kidney disorder given that a patient has, respectively,*

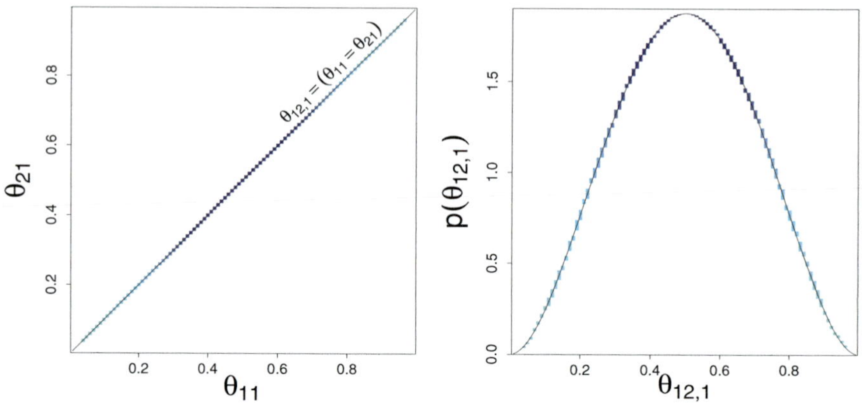

FIGURE 6.3
NLP for the stage u_{12} in the graph $\mathcal{C}_1$ depicted in Figure 6.2b when we set $\overline{\alpha} = 6$. This NLP is equal to a Dirichlet LP given by $\boldsymbol{\theta}_{12} \sim \text{Beta}(3, 3)$. Darker colour represents higher probability densities.

a liver disorder or not. The figure also shows two families of pm-NLPs according to the adopted distance d: here, both Euclidean distance and Hellinger distance [83]. Compare also Theorem 4 below. In Figures 6.4a and 6.4b we observe in particular that the pm-NLPs vanish when these parameters are close to each other, that is when their values are along the diagonal. Since these vanishing regions are around the parameter space of the model represented by $\mathcal{C}_1$ *in Figure 6.3, pm-NLPs only allow the stages* u_1 *and* u_2 *to be identified with each other under the model* $\mathcal{C}_1$*. In doing so, the pm-NLPs inhibit the complex model* $\mathcal{C}_0$ *from representing the same stage structure of the simple model* $\mathcal{C}_1$*, namely* $\boldsymbol{\theta}_1 = \boldsymbol{\theta}_2$.

The pm-NLPs concentrate the prior probability mass in the probability space where the conditional probabilities associated with stages u_1 *and* u_2 *are different. In contrast, Dirichlet LPs set the prior probability mass of the parameters* $\boldsymbol{\theta}_1$ *and* $\boldsymbol{\theta}_2$ *of* $\mathcal{C}_0$ *around the probability parameter space of* $\mathcal{C}_1$*. Therefore, local priors do not ensure a full separation of the parameter space of nested models: the null hypothesis* H_0 *represented by graph* $\mathcal{C}_1$ *is nested into the hypothesis* H_1 *corresponding to graph* $\mathcal{C}_0$*. Also observe that since pm-NLPs embed the distance between models as measured by their nested stages, situations tend to be merged into a single stage in the graph* $\mathcal{C}_1$ *unless their conditional probabilities are acceptably different as in the graph* $\mathcal{C}_0$.

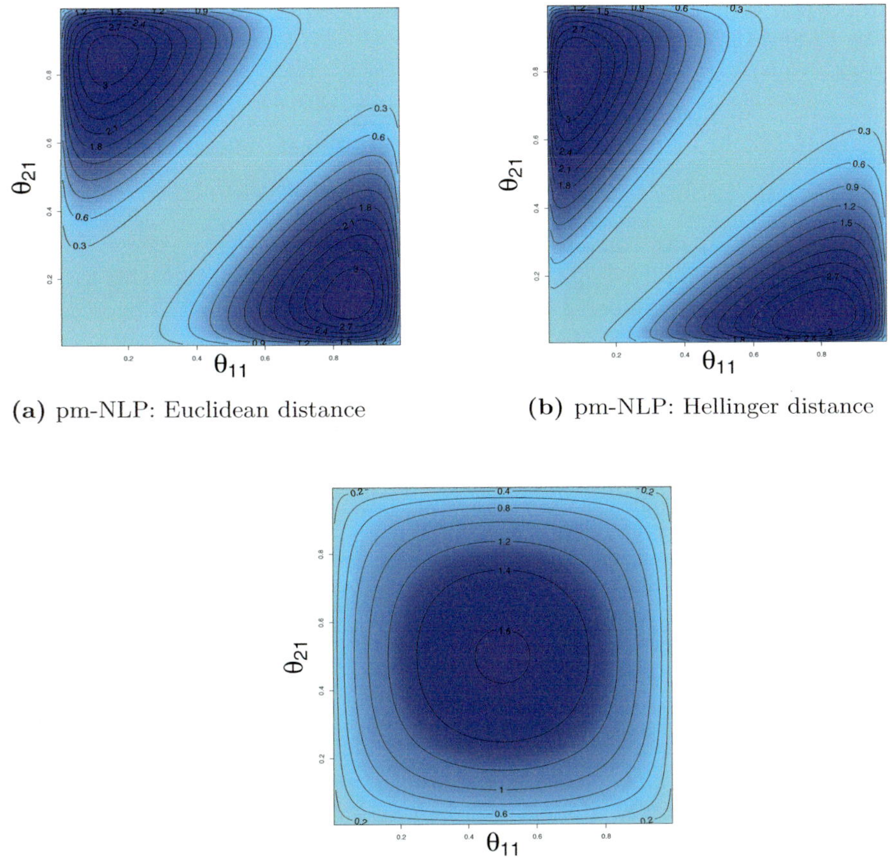

(a) pm-NLP: Euclidean distance

(b) pm-NLP: Hellinger distance

(c) Dirichlet LP

FIGURE 6.4
Dirichlet LP and pm-NLPs using Euclidean and Hellinger distances for stages associated with the variable X_k in the graph $\mathcal{C}_0$ depicted in Figure 6.2a when we set $\overline{\alpha} = 6$. Here, $\boldsymbol{\theta}_1, \boldsymbol{\theta}_2 \sim \text{Beta}(1.5, 1.5)$. Darker colour represents an area of higher joint probability.

Assume now a uniform prior over the space of possible CEGs as in (6.4) and consider again a CEG $\mathcal{C}_i$ and its 1-nested CEG $\mathcal{C}_{i+1}$ where two stages u_1 and u_2 are merged into $u_{12} = u_1 \cup u_2$. Let $\hat{\mathcal{C}}_i$ be the CEG whose graph is identical to $\mathcal{C}_i$ and whose probability measure is based on pm-NLPs defined in relation to $\mathcal{C}_{i+1}$ as in Definition 6.5.

Then the log-posterior Bayes Factor between the two models represented by $\hat{\mathcal{C}}_i$ and $\mathcal{C}_{i+1}$ is equal to

$$\begin{aligned} \mathrm{lpBF}(\hat{\mathcal{C}}_i, \mathcal{C}_{i+1}) &= \log\left(\frac{c_+}{c} \cdot \frac{f_{\mathrm{LP}}(\boldsymbol{y}|\boldsymbol{\theta}_i)}{f_{\mathrm{LP}}(\boldsymbol{y}|\boldsymbol{\theta}_{i+1})} \cdot \frac{q(\mathcal{C}_i)}{q(\mathcal{C}_{i+1})}\right) \\ &= \log c_+ - \log c + \mathrm{lpBF}(\mathcal{C}_i, \mathcal{C}_{i+1}). \end{aligned} \tag{6.14}$$

where c and c_+ are constants.

Theorem 4 explicitly gives the normalisation constants for this formula in closed form when we set $\rho = 1$ and use the Euclidean or Hellinger distances to set up the pm-NLPs of $\hat{\mathcal{C}}_i$.

Theorem 4 (lpBFs for pm-NLPs)**.** *In (6.14) the normalisation constants for $\rho = 1$ are defined according to the distance d used to construct the pm-NLPs of the CEG $\hat{\mathcal{C}}_i$ as follows.*

1. *In Euclidean distance, $c = g(\boldsymbol{\alpha}_1, \boldsymbol{\alpha}_2)$ and $c_+ = g(\boldsymbol{\alpha}_{1+}, \boldsymbol{\alpha}_{2+})$ where*

$$g(\gamma_1, \gamma_2) = \sum_{j=1}^{K_1}\left[\frac{\gamma_{1j}(\gamma_{1j}+1)}{\overline{\gamma}_1(\overline{\gamma}_1+1)} - 2\frac{\gamma_{1j}\gamma_{2j}}{\overline{\gamma}_1\overline{\gamma}_2} + \frac{\gamma_{2j}(\gamma_{2j}+1)}{\overline{\gamma}_2(\overline{\gamma}_2+1)}\right] \tag{6.15}$$

for $(\gamma_1, \gamma_2) = (\boldsymbol{\alpha}_1, \boldsymbol{\alpha}_2), (\boldsymbol{\alpha}_{1+}, \boldsymbol{\alpha}_{2+})$.

2. *In Hellinger distance,*

$$c = 2 - 2\sum_{j=1}^{K_1}\frac{h(\alpha_{1j}, \alpha_{2j})}{h(\overline{\alpha}_1, \overline{\alpha}_2)} \quad \text{and } c_+ = 2 - 2\sum_{j=1}^{K_1}\frac{h(\alpha_{1j+}, \alpha_{2j+})}{h(\overline{\alpha}_{1+}, \overline{\alpha}_{2+})},$$

where $h(\gamma_1, \gamma_2) = \Gamma(\gamma_1+0.5)\Gamma(\gamma_2+0.5)/\Gamma(\gamma_1)\Gamma(\gamma_2)$ for (γ_1, γ_2) as above.

See [17] for a proof of this result.

In this sense, pm-NLPs can be interpreted as imposing a penalty over the complex model using the distance between the conditional probability distributions of stages that are merged in the simple model. It is straightforward to use the pm-NLPs in conjuction with the AHC algorithm since this requires only to add a term $\log c_+ - \log c$ to the standard lpBF score based on Dirichlet conjugate priors. So regardless of their minor global inconsistency discussed above, pm-NLPs enjoy the advantages of NLPs with the extra property that they are also computationally tractable.

We illustrate these results in Section 7.5.2 using a real-world dataset.

6.5 Exercises

Exercise 6.1. *Show using the tools from Chapter 4 that if one CEG is 1-nested in the other then the models represented by these CEGs are submodels of each other. Generalise this concept to k-nested CEGs where k stages are merged.*

Exercise 6.2. *Prove the following. If two different CEG models on the same set of atoms but with different staged trees, under conjugate Dirichlet priors and conserving mass, have different atomic probabilities then their marginal likelihoods will differ for some observations.*

Exercise 6.3. *Suppose that a CEG $\mathcal{C}_2$ is constructed from another CEG $\mathcal{C}_1$ by merging two stages of $\mathcal{C}_2$ into one but otherwise retaining the event tree of $\mathcal{C}_1$ and all its remaining stages, so that $\mathcal{C}_2$ is 1-nested into $\mathcal{C}_1$. Prove that with the calibrated prior defined above the difference of the log-marginal likelihood of $\mathcal{C}_1$ and the log-marginal likelihood of $\mathcal{C}_2$ is a function only of the data and phantom data informing the merged stage. Write down this expression explicitly.*

Exercise 6.4. *Write down the log-posterior Bayes Factor* $\mathrm{lpBF}(\mathcal{C}_1, \mathcal{C}_2)$ *of the models $\boldsymbol{M}_1$ and $\boldsymbol{M}_2$ from Exercise 6.3.*

7

How to model with a CEG: A real-world application

In Chapters 5 and 6 we have built the algorithmic foundations which enable us to now discuss various features of CEG modelling using a real-world example. We will start from introducing our dataset, discuss how domain knowledge can be embedded in possible CEG models for this dataset and decide how to set up a prior distribution over these models. We will then run the algorithms introduced in the previous chapter in our R-package `ceg` [18]. The reader will be asked to redo these calculations in an exercise at the end of this chapter. Throughout, we discuss the fit of the different CEG models we discover and compare their respective interpretation in terms of the study which produced the data.

The dataset we consider here constitutes a small part of the data collected for the *Christchurch Health and Development Study (CHDS)* carried out at the University of Otago, New Zealand [14]. We will henceforth refer to these data as the *CHDS data*. The CHDS is a 5-year longitudinal study of rates of childhood hospitalisation based on a cohort of 1,265 children born in 1977. The children's family were interviewed at birth, four months after birth and once in each of the following years until their child reached the age of five. Information on each child's development was collected in four different ways: a structured interview with the child's mother, a diary completed by the child's mother, hospital records, and practitioner notes [30–32].

For the purpose of our analyses we model the rates of hospital admission of a child as a function of the following three discrete explanatory variables:

Social background X_s: a categorical variable distinguishing between high (h) and low (l) levels. This variable was constructed using a latent-class model based on measures of maternal educational level and age at the child's birth, the child's ethnicity, the family's social class and whether a child entered an adoptive, a single or two-parent family.

Economic status X_e: a categorical variable differentiating between high (h) and low (l) status. This variable was obtained from a latent-class model whose inputs were income, standard of living, financial difficulty and the quality of the accommodation inhabited by the family of each child.

Life events X_l: a categorical variable indicating whether the family of a child

TABLE 7.1
Summary statistics of the CHDS dataset for the variables social status, economic situation and life events against hospital admission.

Hospital Admission	Social Status		Economic Status		Life Events			Total
	Low	High	Low	High	Low	Moderate	High	
No	289	432	480	241	290	233	198	721
Yes	94	75	127	42	39	62	68	169
Total	383	507	607	283	329	295	266	890

experiences a low (0 to 5 events; l), moderate (6 to 9 events; m) or high (10 or more events; h) number of stressful events over the monitored period of five years. This includes events such as death, illness, unemployment and marital disharmony.

One of the diverse objectives of this study was to explore how social and economic factors associated with the stress faced by a family can affect the risk of hospitalisation during childhood. **Hospital admission** X_{h} will thus be our response variable which is assumed to be binary with levels no (n) and yes (y) and signals whether a child was hospitalised at least one time during his first five years of life. In this study, only hospitalisations for respiratory infections, gastroenteritides and accidents were considered. The CHDS dataset available to us has a complete record of 890 children. The corresponding summary statistics are presented in Table 7.1. For a detailed description of the collection and pre-processing of this dataset see [5] and [32].

In Section 7.1 we first revisit two previous studies on this dataset conducted by the group who originally designed the CHDS and show how their results can alternatively be represented using the CEG framework. In Sections 7.2 and 7.3 we will then go on to discuss how a CEG model can further refine a well-fitted BN model using, respectively, the AHC algorithm—Algorithm 2 from Section 6.3.1—and the exhaustive model search algorithm—Algorithm 4 described in Section 6.3.2. These results will then further be compared with CEG model selection techniques using pm-NLPs as in Section 6.4.2. Finally in Section 7.4 we use the DP programming algorithm—Algorithm 8 from Section 6.4.1—to search the CEG model space with a block ordering and without a variable ordering. Concluding this chapter, a number of challenges will be discussed in Section 7.5.

7.1 Previous studies and domain knowledge

Using the information provided in the previous section, a set of four random variables which we will throughout denote as $\boldsymbol{\mathcal{X}} = \{X_{\mathrm{s}}, X_{\mathrm{e}}, X_{\mathrm{l}}, X_{\mathrm{h}}\}$ defines the hospitalisation process described in the CHDS. So the corresponding CEG model space $\boldsymbol{C}(\boldsymbol{\mathcal{X}})$ will be spanned by a set of $4! = 24$ $\boldsymbol{\mathcal{X}}$-compatible event trees, each of which corresponds to a possible permutation of these variables. Since the aim is to explore the impact of the three explanatory variables—social and economic status and life events—on childhood hospitalisation we have a block ordering $\boldsymbol{\mathcal{B}} = \{\boldsymbol{\mathcal{B}}_1, \boldsymbol{\mathcal{B}}_2\}$ of this set of measurement variables where $\boldsymbol{\mathcal{B}}_1 = \{X_{\mathrm{s}}, X_{\mathrm{e}}, X_{\mathrm{l}}\}$ and $\boldsymbol{\mathcal{B}}_2 = \{X_{\mathrm{h}}\}$. Under this ordering, the set of event trees $\mathcal{T}_{\boldsymbol{\mathcal{X}}}$ which span the CEG model space is reduced to six $\boldsymbol{\mathcal{X}}$-compatible event trees associated with the unknown variable ordering within the block $\boldsymbol{\mathcal{B}}_1$.

Two studies conducted by researchers in the CHDS group make more domain information available. The first work carried out by Fergusson et al. [30] in 1984 analysed the influence of social background, family economic situation, family adversity and family composition on the utilisation of preschool health and education services based on information on 1,080 children collected in the CHDS dataset. All of these variables were obtained using a factor analysis. The first two variables closely resemble the construction of variables X_{s} and X_{e} in terms of observed indicator variables and will hence be denoted X_{s_1} and X_{e_1}, respectively. A new variable *family adversity* X_{a} extends the concept of family live events by including, for example, the number of changes of residence and maternal depression. Finally, the variable *family composition* X_{c} is a new measure which captures the birth order of a child, the number of children that the family had after the study child and whether the pregnancy was planned.

Fergusson and his coauthors found a strong correlation between each of the above explanatory variables and *service utilisation* X_{u} without adjusting for other covariates. They also constructed a hierarchical linear model whose structural equations assumed the variable ordering $\boldsymbol{X}(I) = (X_{\mathrm{s}_1}, X_{\mathrm{c}}, X_{\mathrm{e}_1}, X_{\mathrm{a}}, X_{\mathrm{u}})$. Their results pointed out again a strong total correlation between each explanatory variable and the rates of utilisation of care. A strong mutual correlation between the variables X_{s_1}, X_{e_1} and X_{a} was also observed, particularly between the variables X_{s_1} and X_{e_1}.

The second study, this time conducted by Fergusson et al. [32] in 1986, was the first work to analyse the relation between the set of social, economic and stress factors and the rate of childhood admission using a subset of 1,057 children collected in the CHDS dataset. For this purpose, the authors constructed again the variables social background X_{s_2} and economic situation X_{e_2} using factor analysis. Life events constituted a third explanatory variable $X_{\mathrm{l}(t)}$ which was assumed to vary each year $t = 2, 3, 4, 5$. A one-way analysis of variance between each of these covariates and the rate of hospitalisation

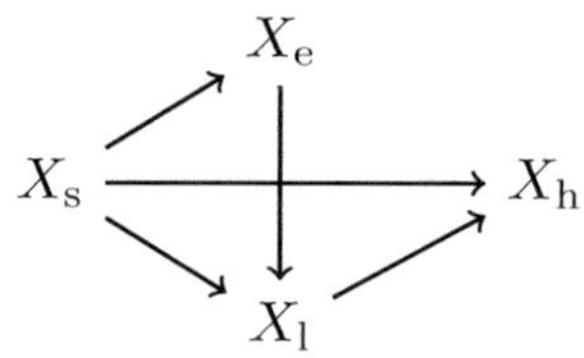

FIGURE 7.1
A DAG representation for the **BN A** obtained from previous studies using the CHDS dataset [30, 32].

indicated that the likelihood of hospitalisation increases among children who face social disadvantages, depressed material condition and family stress.

For further analysis the authors fitted a Cox proportional hazards model to model a child's risk of at least one hospital admission during his first five years of life, denoted X_{h_2}. The explanatory variables used here were the time dependent covariate $X_{l(t)}$ and the time-invariant covariates X_{s_2} and X_{e_2}. The authors concluded that the variables social status X_{s_2} and life events $X_{l(t)}$ had a significant influence on hospital admission whilst the economic status X_{e_2} was not a significant predictor. This model assumed an implicit block ordering given by $\{X_{s_2}, X_{e_2}\}$, $\{X_{l(t)}\}$ and $\{X_{h_2}\}$.

From these two previous studies we can assume that domain experts would expect a variable ordering given by $\boldsymbol{X}(I_1) = (X_s, X_e, X_l, X_h)$ and that there is a single conditional independence statement given by $X_h \perp\!\!\!\perp X_e \mid X_s, X_l$. Figure 7.1 depicts the only possible BN—we will call this BN A—that fully represents this information. The score given by the log-marginal likelihood of the model $\boldsymbol{M}_{\text{BN A}}$ represented by the DAG in the figure above is $Q(\boldsymbol{M}_{\text{BN A}}) = -2,495.01$.

This BN model provides us with a very concise representation of the information available. Note that this is not the case for the alternative CEG representation of this model whose event tree is depicted in Figure 7.2. This is because there is only one symmetric conditional independence statement. As a consequence, we can only identify situations corresponding to the variable X_h and we need to use six colours to depict this information. This implies that the CEG model has 14 distinct positions which will not allow for a very compact representation of the model. As discussed in Chapter 3, in fact the CEG framework is at its most useful when a process has asymmetric developments and/or is driven by context-specific conditional structures.

However, a very different picture emerges after learning the BN model depicted in Figure 7.1. For this purpose, we assume a weak Dirichlet informative prior by setting the equivalent sample size hyperparameter to $\overline{\alpha} = 3$ and then distributing it uniformly over the network. This value was chosen because it corresponds to the maximum number of categories that a variable in $\boldsymbol{\mathcal{X}}$ can take: see the discussion in Section 5.1.2 and the recommendation in [72]. A careful examination of the mean posterior conditional probability table asso-

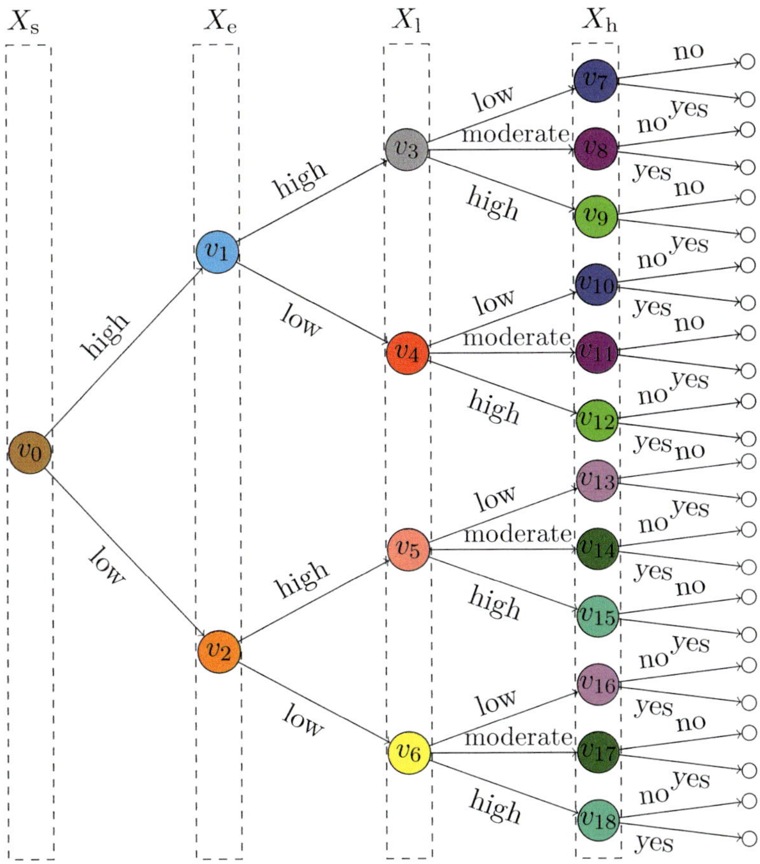

FIGURE 7.2
The $\boldsymbol{X}(I_1)$-compatible staged tree representing the **BN A** depicted in Figure 7.1 obtained from previous studies using the CHDS dataset [30, 32].

ciated with this BN as given in Table 7.2 enables us to conclude that there may well be some context-specific conditional independence statements that are kept hidden. In particular, we observe the following:

1. The mean posterior probability of stressful events in a socially advantaged but economically deprived family is very close to the one in a family facing social problems but having a comfortable economic situation: see the red-coloured cells in Table 7.2. This implies that we should assign the same colour to the situations v_4 and v_5 in Figure 7.2.
2. The mean posterior risk of hospitalisation of a child from a socially disadvantaged family is independent from the number of family life events given that this number is not low. The mean posterior probability of hospital-

TABLE 7.2
Mean posterior conditional probability table for the **BN A** depicted in Figure 7.1 using the CHDS dataset with $\overline{\alpha} = 3$.

Variable	Conditional probability vector	Stage
X_s	$P(X_s = (h,l)) = (0.57,0.43)$	
X_e	$P(X_e = (h,l) \mid X_s = h) = (0.47,0.53)$	
	$P(X_e = (h,l) \mid X_s = l) = (0.12,0.88)$	
X_l	$P(X_l = (h,m,l) \mid X_s = h, X_e = h) = (0.14,0.36,0.50)$	
	$P(X_l = (h,m,l) \mid X_s = h, X_e = l) = (0.24,0.33,0.43)$	
	$P(X_l = (h,m,l) \mid X_s = l, X_e = h) = (0.24,0.31,0.45)$	
	$P(X_l = (h,m,l) \mid X_s = l, X_e = l) = (0.47,0.31,0.22)$	
X_h	$P(X_h = (n,y) \mid X_s = h, X_l = h) = (0.75,0.25)$	
	$P(X_h = (n,y) \mid X_s = h, X_l = m) = (0.83,0.17)$	
	$P(X_h = (n,y) \mid X_s = h, X_l = l) = (0.91,0.09)$	
	$P(X_h = (n,y) \mid X_s = l, X_l = h) = (0.74,0.26)$	
	$P(X_h = (n,y) \mid X_s = l, X_l = m) = (0.73,0.27)$	
	$P(X_h = (n,y) \mid X_s = l, X_l = l) = (0.81,0.19)$	

isation of a child is also independent of his family social background if his family experiences a high number of stressful life events. We indicate these context-specific conditional independences using green cells in Table 7.2. These assumptions thus translate into a shared colouring of all situations in $v_9, v_{12}, v_{14}, v_{15}, v_{17}, v_{18}$ in Figure 7.2.

3. The mean posterior likelihood of hospital admission of a socially advantaged child does not depend on the number of negative life events experienced by his family if this number is not moderate. We colour the corresponding cells violet in Table 7.2. This now also implies that the situations v_8, v_{11}, v_{13} and v_{16} in Figure 7.2 should be given the same colour.

These conclusions cannot be represented through a DAG. However, they are easily depicted using the CEG model $\boldsymbol{M}_{\text{CEG A}}$ represented in Figure 7.3. We are now able to obtain a very concise model representation. From this, the domain experts can easily and directly read the qualitative conditional independences, without consulting the conditional probability tables. This happens because all qualitative information is directly embedded into the CEG using colours and labelling the edges with the variable categories.

Table 7.3 shows the mean posterior probability distribution associated with the CEG depicted in Figure 7.3 when we learn it using the CHDS dataset. In order to enable a direct comparison with the BN model, we set a weakly

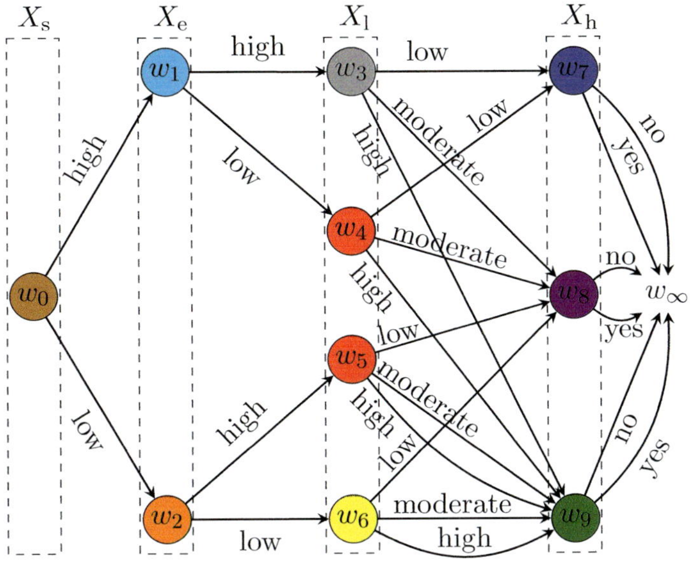

FIGURE 7.3
The **CEG A** improving the BN A obtained from previous studies using the CHDS dataset [30, 32]. The stage structure is given by: $u_0 = \{w_0\}$, $u_1 = \{w_1\}$, $u_2 = \{w_2\}$, $u_3 = \{w_3\}$, $u_4 = \{w_4, w_5\}$, $u_5 = \{w_6\}$, $u_6 = \{w_7\}$, $u_7 = \{w_8\}$, $u_8 = \{w_9\}$. The colours of this graph have been transferred from the staged tree in Figure 7.2 and Table 7.2.

informative prior using a hyperparameter $\overline{\alpha} = 3$ distributed uniformly over the graph. This table has far less cells than Table 7.2. As discussed in Section 5.2.1, this implies that collecting and propagating evidence over the CEG model is computationally more efficient than using its corresponding BN model. The score given by the log-marginal likelihood of this model is $Q(\boldsymbol{M}_{\text{CEG A}}) = -2,480.31$. Assuming a uniform prior over the model space, (6.2) implies that this CEG model therefore has a Bayes Factor of

$$\text{BF}(\boldsymbol{M}_{\text{CEG A}}, \boldsymbol{M}_{\text{BN A}}) = \exp(-2,480.31 + 2,495.01) = 2,421,748 \tag{7.1}$$

in its favour when it is compared with the BN model. Thanks to the implicit conditional-independence information, we have thus found a CEG model which is more than two million times more likely to have generated the CHDS data than the BN drawn out from previous studies. This is extreme evidence for the CEG rather than the BN describing the data-generating process behind the CHDS dataset [5].

Now the proposed CEG model enables us further to draw out even more information about the children in the study. In particular, we can directly identify three groups of families classified according their likelihood of experiencing stressful events. The first group (in stage u_3) is constituted by famil-

TABLE 7.3
Mean posterior conditional probability table for the **CEG A** depicted in Figure 7.3 using the CHDS dataset with $\overline{\alpha} = 3$.

Variable	Conditional probability vector
X_s	$P(X_s = (h, l) \mid u_0) = (0.57, 0.43)$
X_e	$P(X_e = (h, l) \mid u_1) = (0.47, 0.53)$
	$P(X_e = (h, l) \mid u_2) = (0.12, 0.88)$
X_l	$P(X_l = (h, m, l) \mid u_3) = (0.14, 0.36, 0.50)$
	$P(X_l = (h, m, l) \mid u_4) = (0.24, 0.33, 0.43)$
	$P(X_l = (h, m, l) \mid u_5) = (0.47, 0.31, 0.22)$
X_h	$P(X_h = (n, y) \mid u_7) = (0.91, 0.09)$
	$P(X_h = (n, y) \mid u_8) = (0.82, 0.18)$
	$P(X_h = (n, y) \mid u_9) = (0.74, 0.26)$

ies that enjoy a socially and economically comfortable situation. The second group (in stage u_4) is formed by socially advantaged families with a prosperous economic status and economically deprived families with a good social background. The last group (in stage u_5) consisted of socially disadvantaged and economically depressed families. As expected, the risk of experiencing adverse events increases monotonically from the first group to the last one.

In terms of childhood hospital admission, we can also divide the CHDS children into three groups. Children from socially advantaged families that face few adversities (in stage u_6) are less prone to be hospitalised. Children from socially advantaged but moderately stressed families and children from socially disadvantaged but non-stressed families (in stage u_7) have an intermediate likelihood of being admitted at hospital. Finally, children living in a highly stressed family and children from low social status families that face a moderate number of adverse events (in stage u_8) constitute the group which is most at risk.

7.2 Searching the CHDS dataset with a variable order

Based on the analysis in Section 7.1 we can expect that the CHDS data-generating process is driven by some context-specific structures rather than by conditional independences between the pre-determined random variables in the set $\boldsymbol{\mathcal{X}}$. The two analytic methods used previously have not been able to fully handle the conditional asymmetries that may be presented in the data

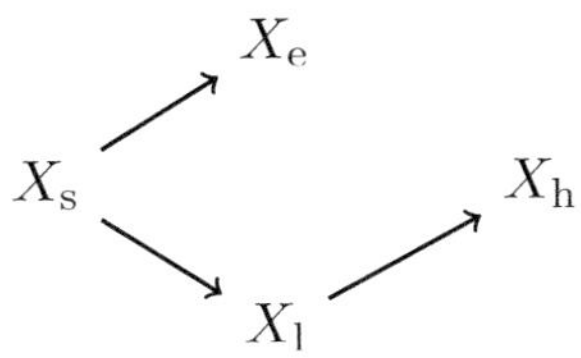

FIGURE 7.4
A DAG for the MAP **BN B** obtained using the CHDS dataset when we fix the variable order $\boldsymbol{X}(I_1) = (X_s, X_e, X_l, X_h)$ and set $\overline{\alpha} = 3$. Compare also Figure 7.1.

and neither is the BN framework. However, being a particular subfamily of CEGs, the class of BN models provides us with a sound and reliable graphical background to analyse the results when we search the CEG model space: compare Section 6.3. For this purpose, we will now first assume the variable order $\boldsymbol{X}(I_1) = (X_s, X_e, X_l, X_h)$ above and always fix the equivalent sample size at $\overline{\alpha} = 3$.

Figure 7.4 depicts the MAP BN—here referred to as BN B—when an exhaustive model search for the chosen variable ordering is performed using Bayes Factor scores as discussed in Section 2.1.4. In contrast to the previous BN A from Figure 7.1, here there are two more conditional independence statements, namely: $X_h \perp\!\!\!\perp X_s, X_e \mid X_l$ and $X_l \perp\!\!\!\perp X_e \mid X_s$. Table 7.4 shows the conditional probability tables for this BN model $\boldsymbol{M}_{\text{BN B}}$.

According to this model, we can now distinguish two types of families in the CHDS: the socially advantaged ones that have a low chance of facing adversities and the socially deprived ones that have a high likelihood of experiencing stress. Moreover, children can then be split into three groups according to the stress level faced by their families. In this case, the risk of hospitalisation increases monotonically with the number of adverse events handled by their families.

Although the BN B provides a good level of sparsity that facilitates its interpretation, it appears to be an over-simplification of the data-generating mechanisms. For example, the family socio-economic status does not improve the capability of the number of stressful events to explain the risk of hospitalisation. Also a family's economic background does not bring any additional information to infer the number of life events once the family's social situation is known. These two hypotheses are not usually expected by domain experts and are not often found in the specialised literature. In fact, this model has a score of $Q(\boldsymbol{M}_{\text{BN B}}) = -2,489.78$ which is greater than the first BN A in Figure 7.1 but is substantially lower than the score of the CEG A in Figure 7.3. In this case, the CEG has a BF score of $\text{BF}(\boldsymbol{M}_{\text{CEG A}}, \boldsymbol{M}_{\text{BN B}}) = 12,965$ in its favour. This is strong evidence for the CEG A to represent the more likely data-generating process and to, in this sense, be the better model.

We now run the AHC and the model search algorithms, Algorithms 2

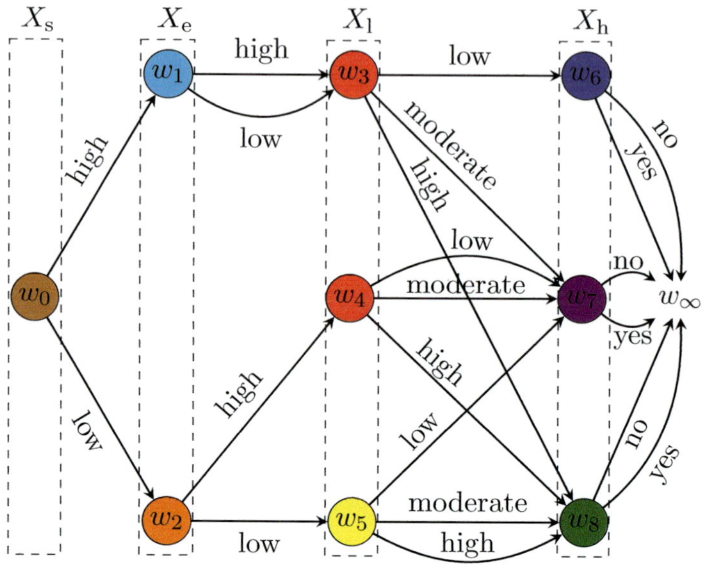

(a) The best **CEG B** found by the AHC algorithm.

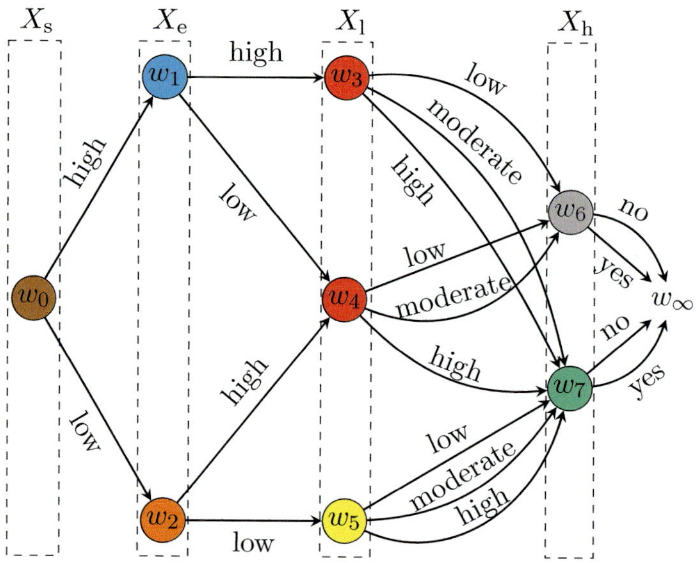

(b) The MAP **CEG C**.

FIGURE 7.5
CEGs found by the AHC and an exhaustive model search algorithm using the CHDS dataset when we fix the variable order $\boldsymbol{X}(I_1) = (X_s, X_e, X_l, X_h)$ and set $\overline{\alpha} = 3$. The stage structure of these graphs is: $u_0 = \{w_0\}$, $u_1 = \{w_1\}$, $u_2 = \{w_2\}$, $u_3 = \{w_3, w_4\}$, $u_4 = \{w_5\}$, $u_5 = \{w_6\}$, $u_6 = \{w_7\}$, $u_7 = \{w_8\}$.

TABLE 7.4
Mean posterior conditional probability table for the **BN B** depicted in Figure 7.4 using the CHDS dataset with $\overline{\alpha} = 3$.

Variable	Conditional probability vector
X_s	$P(X_\text{s} = (h, l)) = (0.57,0.43)$
X_e	$P(X_\text{e} = (h, l) \mid X_\text{s} = h) = (0.47,0.53)$
	$P(X_\text{e} = (h, l) \mid X_\text{s} = l) = (0.12,0.88)$
X_l	$P(X_\text{l} = (h, m, l) \mid X_\text{s} = h) = (0.44,0.31,0.25)$
	$P(X_\text{l} = (h, m, l) \mid X_\text{s} = l) = (0.19,0.35,0.46)$
X_h	$P(X_\text{h} = (n, y) \mid X_\text{l} = h) = (0.74,0.26)$
	$P(X_\text{h} = (n, y) \mid X_\text{l} = m) = (0.79,0.21)$
	$P(X_\text{h} = (n, y) \mid X_\text{l} = l) = (0.88,0.12)$

and 4, from the previous chapter on the CHDS dataset to see whether or not these will find even better scoring models than the MAP BN B or the models based on previous analyses of this study.

Figures 7.5a and 7.5b depict the CEG selected by the AHC algorithm—hence denoted as CEG B—and the MAP CEG—hence denoted CEG C—found by the exhaustive model search algorithm, respectively. Table 7.5 presents the conditional probability tables of these two models. These models have a very similar score: $Q(\boldsymbol{M}_{\text{CEG B}}) = -2,478.49$ for CEG B and $Q(\boldsymbol{M}_{\text{CEG C}}) = -2,478.17$ for CEG C. So the MAP CEG C represents only a slightly BF improvement of $\text{BF}(\boldsymbol{M}_{\text{CEG C}}, \boldsymbol{M}_{\text{CEG B}}) = 1.37$ in comparison with CEG B chosen by the greedy search algorithm. This is very weak evidence in support of model represented by the CEG C in Figure 7.5b. However, both models provide clearly far better fits than the MAP BN B and have a significantly better score than the CEG A. For example, CEG B has a Bayes Factor of $\text{BF}(\boldsymbol{M}_{\text{CEG B}}, \boldsymbol{M}_{\text{CEG A}}) = 6.17$ and $\text{BF}(\boldsymbol{M}_{\text{CEG B}}, \boldsymbol{M}_{\text{BN B}}) = 80,017.45$ compared to CEG A and the MAP BN B, respectively. These models are also more readable than CEG A since they are sparser.

We can now compare these two new models. Note first that both CEGs B and C have the same graphical and conditional probability structures for the three explanatory variables X_s, X_e and X_l. Here, families are divided into two groups according to their chance of facing stressful events. The probabilities of moderate numbers of events in families within these groups are almost identical at 0.34 and 0.31, respectively. The probability of a low or high number of life events are inverse within these groups: for families with lower risk the probability of low and high number of adverse events are, respectively, 0.20 and 0.46, whilst for families with higher risk these probabilities are, respectively, 0.47 and 0.22: compare Table 7.5.

TABLE 7.5
Mean posterior conditional probability table for the **CEGs B and C** depicted in Figure 7.5 using the CHDS dataset with $\overline{\alpha} = 3$.

Variable	Conditional probability vector		CEG B	CEG C
X_s	$P(X_s = (h,l) \mid u_0)$	=	(0.57,0.43)	(0.57,0.43)
X_e	$P(X_e = (h,l) \mid u_1)$	=	(0.47,0.53)	(0.47,0.53)
	$P(X_e = (h,l) \mid u_2)$	=	(0.12,0.88)	(0.12,0.88)
X_l	$P(X_l = (h,m,l) \mid u_3)$	=	(0.20,0.34,0.46)	(0.20,0.34,0.46)
	$P(X_l = (h,m,l) \mid u_4)$	=	(0.47,0.31,0.22)	(0.47,0.31,0.22)
X_h	$P(X_h = (n,y) \mid u_5)$	=	(0.91,0.09)	(0.89,0.11)
	$P(X_h = (n,y) \mid u_6)$	=	(0.82,0.18)	(0.76,0.24)
	$P(X_h = (n,y) \mid u_7)$	=	(0.74,0.26)	—

Also note that now a family's socio-economic status plays an important role in identifying to which of these two groups the family belongs. In fact, socially and economically disadvantaged families constitute the highest risk group of families in terms of their numbers of life events. This is represented by the red stage u_3, which merges the positions w_3 and w_4. The analysis of the low risk families implies that the variable X_l is independent from variables X_s and X_e given that X_s and X_e do not simultaneously assume the value *low*.

The two CEGs in Figure 7.5 differ only on their last level. With regard to childhood hospitalisation, CEG B identifies three risk groups of children whilst CEG C recognises only two groups. The highest and lowest risk groups in CEG B have similar probabilities of hospital admission to their corresponding groups in CEG C: see again Table 7.5. The main difference between these two models is that CEG B recognises an intermediate risk group constituted of children from unstressed families with low social status and also of children from economically advantaged families with a history of moderate number of life events.

Of course, when choosing between these two models we should consider the scope of our analysis. For instance, if these models were designed to support decision makers who face severe financial budget constraints to assist the most vulnerable children in a particular community then CEG B may look more appealing since it provides a finer partition of the children in the population. On the other hand, if the objective of decision makers is to design more inclusive policies that reduce the pressure over the medical facilities, then CEG C may provide a more attractive explanation because it provides us with a clear picture of a small group of children who should not be targeted. In Chapter 8, we will provide the mathematical framework to analyse and perform these types of interventions in a group of children at risk: see Section 8.3, in particular.

Finally, with respect to an interpretation of the chosen model, in CEG C there is a technicality that should be considered by analysts and domain experts. This is that the lowest risk group includes the following sub-populations:

1. children from unstressed families with a high social-economic situation;
2. children from socially advantaged but economically deprived families facing a low or a moderate number of adverse life events; and
3. children from economically advantaged but socially depressed families experiencing a low or a moderate number of stressful events.

So children from both socially and economically advantaged families with a history of only a moderate amount of life events are part of the highest risk group! This conclusion can be interpreted as an inconsistency due to sampling effect and/or model selection bias. In fact, it cannot reasonably be expected that a child from a family with a comfortable socio-economic background will have a higher risk of hospitalisation than a child with a lower socio-economic status if both children live in families under the same level of stress. Observe that however these domain hypotheses hold and are not contradictory in CEG B.

7.3 Searching the CHDS dataset with a block ordering

Now we proceed to analyse the CHDS data without considering a fixed variable order $\boldsymbol{X}(I_1)$ over the explanatory socio-economic and life event variables in the set $\boldsymbol{\mathcal{X}}$ as above. Here, we will only assume to have the block ordering $\boldsymbol{\mathcal{B}} = \{\boldsymbol{\mathcal{B}}_1, \boldsymbol{\mathcal{B}}_2\}$ into explanatory variables $\boldsymbol{\mathcal{B}}_1 = \{X_{\text{s}}, X_{\text{e}}, X_{\text{l}}\}$ and the response variable $\boldsymbol{\mathcal{B}}_2 = \{X_{\text{h}}\}$. Under this assumption, we will again compare the results provided by a greedy search algorithm with those obtained using an exhaustive model selection algorithm.

Recall that the CEG model space associated with the CHDS dataset is spanned by $\boldsymbol{\mathcal{X}}$-compatible event trees and so its elements are stratified CEG models which embellish somewhat the corresponding BN models. This enables us to adopt the greedy model search described in Section 6.3.1 that first looks for the MAP BN model and then refines it using the AHC algorithm presented in Algorithm 2. In this case, the MAP BN provides us with a variable order to be used by the AHC algorithm.

One of the MAP BNs we can derive from the CHDS dataset is the BN B which has already been represented by the DAG in Figure 7.4. This particular DAG gives us the two possible variable orderings $\boldsymbol{X}(I_1) = (X_{\text{s}}, X_{\text{e}}, X_{\text{l}}, X_{\text{h}})$—analysed above—and $\boldsymbol{X}(I_2) = (X_{\text{s}}, X_{\text{l}}, X_{\text{e}}, X_{\text{h}})$. However, we also have to consider its statistically equivalent model representations. Figure 7.6 depicts the other two possible DAG representations of this MAP BN under the block ordering $\boldsymbol{\mathcal{B}}$ given above. The BN models $\boldsymbol{M}_{\text{BN C}}$ represented in Figure 7.6a and $\boldsymbol{M}_{\text{BN D}}$ represented in Figure 7.6b are equal but induce, respectively, two new

(a) A DAG of the **BN C**. **(b)** A DAG of the **BN D**.

FIGURE 7.6
Two MAP BNs obtained using the CHDS dataset when we assume the block ordering $\boldsymbol{\mathcal{B}} = \{\boldsymbol{\mathcal{B}}_1, \boldsymbol{\mathcal{B}}_2\}$, where $\boldsymbol{\mathcal{B}}_1 = \{X_s, X_e, X_l\}$ and $\boldsymbol{\mathcal{B}}_2 = \{X_h\}$, and set again $\overline{\alpha} = 3$. Of course, the BNs B, C and D are statistically equivalent, so the DAGs depicted here and in Figure 7.4 represent the same model.

variable orderings $\boldsymbol{X}(I_3) = (X_e, X_s, X_l, X_h)$ and $\boldsymbol{X}(I_4) = (X_l, X_s, X_e, X_h)$. Now in order to proceed with our analysis we will have to select one of these four possible orders of variable. Of course, the natural choice would be to pick up $\boldsymbol{X}(I_1)$ but this alternative was exhaustively discussed previously.

Another natural choice is the variable ordering $\boldsymbol{X}(I_3)$. Compared to $\boldsymbol{X}(I_1)$, this variable ordering only permutes the first two explanatory variables X_s and X_e. This appears reasonable in that these have both a similar expected role in the description of this process. Moreover, domain experts often assume that social-economic status tend to be more robust to change than the number of life events. For this reason it is more compelling to explain the level of family stress using the social and economic variables than to look for other explanatory mechanisms. This assumption is consistent with the block ordering of the Cox proportional hazards model discussed in Section 7.1.

A technical reason for adopting the variable ordering $\boldsymbol{X}(I_3)$ rather than $\boldsymbol{X}(I_2)$ or $\boldsymbol{X}(I_4)$ is that the corresponding $\boldsymbol{X}(I_3)$-compatible event tree $\mathcal{T}(\boldsymbol{X}(I_3))$ is more compact, in that is has a smaller number of edges and hence of free parameters, than those corresponding to $\mathcal{T}(\boldsymbol{X}(I_2))$ or $\mathcal{T}(\boldsymbol{X}(I_4))$. This happens because the variable X_l has three categories whilst the social and economic variables are binary. So the number of situations associated with the first levels of $\mathcal{T}(\boldsymbol{X}(I_2))$ or $\mathcal{T}(\boldsymbol{X}(I_4))$ is higher than those associated with $\mathcal{T}(\boldsymbol{X}(I_3))$. We can see this in Figure 7.7 which shows the event tree $\mathcal{T}(\boldsymbol{X}(I_4))$. This is the most complex event tree in the CHDS CEG model space since the variable X_l, which has the largest number of categories, is associated with the sink node. However, this additional complexity does still not enable the model search algorithm to discover further explanatory structures for childhood hospitalisation because these structures are independent of the explanatory variable ordering. We will thus start off with analysing the stratified CEG model subclass spanned by $\boldsymbol{X}(I_3)$-compatible event trees in $\mathcal{T}_{\boldsymbol{X}}$.

Figure 7.8 depicts the highest scored CEG—henceforth called CEG D—found by the AHC algorithm when we assume the variable ordering $\boldsymbol{X}(I_3)$ and fix $\overline{\alpha} = 3$. The stage colours here are equivalent to those used in CEG B in the

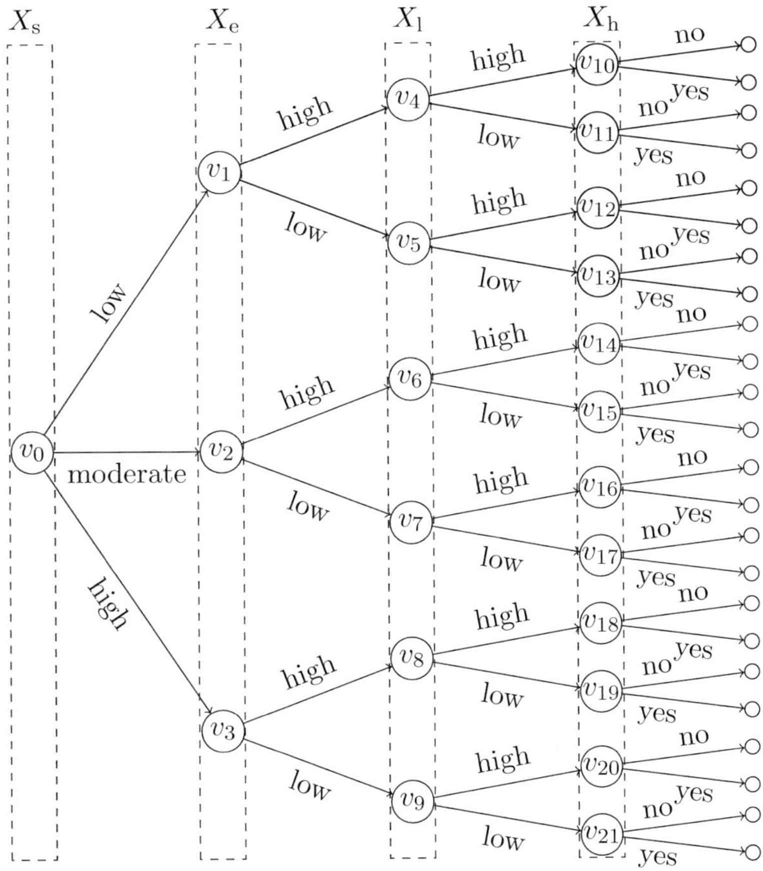

FIGURE 7.7
An event tree for the CHDS dataset when the alternative variable ordering $\boldsymbol{X}(I_4) = (X_l, X_s, X_e, X_h)$ is assumed.

sense that stages which are coloured the same in Figures 7.5a and 7.8 collect together completely analogous situations in the underlying event trees of both models. For instance, in both models the stage u_4 represents families with low socio-economic status. In contrast, the stage u_1 corresponds to a socially advantaged family in CEG B but to an economically advantaged family in CEG D. The conditional probabilities for this new model $\boldsymbol{M}_{\text{CEG D}}$ are shown in Table 7.6.

It is straightforward to verify that the CEG D is statistically equivalent to the CEG B depicted in Figure 7.5a. In other words, the corresponding models are the same, $\boldsymbol{M}_{\text{CEG B}} = \boldsymbol{M}_{\text{CEG D}}$. This is simply because they entail the same conditional independence structures: compare Tables 7.5 and 7.6. In fact, it would not be even necessary to run the AHC algorithm to infer

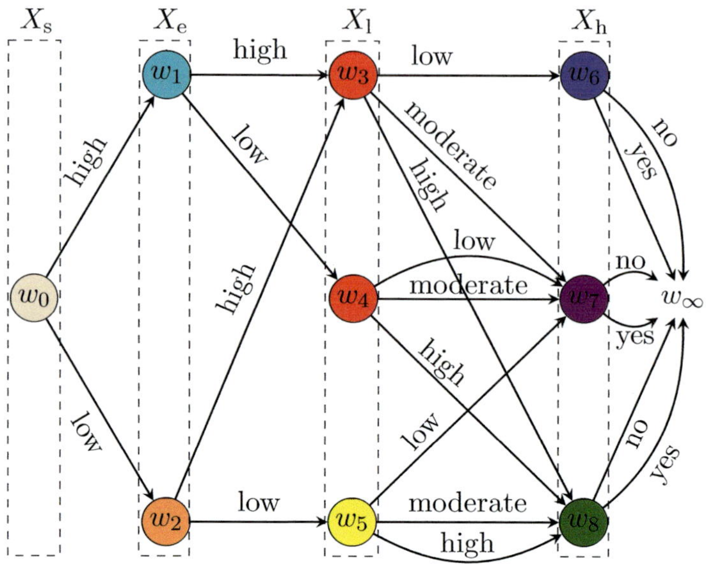

FIGURE 7.8
The **CEG D** found by the AHC algorithm using the CHDS dataset when we fix the variable order $\boldsymbol{X}(I_3) = (X_e, X_s, X_l, X_h)$ and set $\overline{\alpha} = 3$. The stage structure is given by: $u_0 = \{w_0\}$, $u_1 = \{w_1\}$, $u_2 = \{w_2\}$, $u_3 = \{w_3, w_4\}$, $u_4 = \{w_5\}$, $u_5 = \{w_6\}$, $u_6 = \{w_7\}$, $u_7 = \{w_8\}$. Colours are used in analogy to those used in Figure 7.5a.

that the CEG D would be selected given the variable ordering $\boldsymbol{X}(I_3)$. To see this recall that the best stage configuration associated with variables X_l and X_h does not depend on the variable ordering of their preceding variables: see Section 6.3.2. Thus, the stage configurations for these variables are necessarily identical regardless of which of the two variable orderings $\boldsymbol{X}(I_1)$ or $\boldsymbol{X}(I_3)$ is used. Finally, since the variables X_s and X_e are binary and are not independent according to CEG B, the stage structure with two different stages associated to the variable X_s in CEG D is the only valid one. Otherwise, CEG D would imply that the variables X_e and X_s are independent and this would contradict CEG B. In terms of the theory developed in Section 4.2, CEG B and CEG D are statistically equivalent because the first two levels of these CEGs can be resized into the same floret, depicting a joint random variable 'socio-economic status' with no (context-specific) conditional independence information.

To further explore the CHDS dataset we can also examine the two remaining variable orderings $\boldsymbol{X}(I_2)$ and $\boldsymbol{X}(I_4)$. Figure 7.9 shows the highest scored CEGs E and F found by the AHC algorithm when we assume one of these variable orders, respectively. These models are again statistically equivalent—the first two levels of these CEGs are saturated with the same probability distributions and are hence resizeable—and they have the same

TABLE 7.6
Mean posterior conditional probability table for the **CEG D** depicted in Figure 7.8 using the CHDS dataset with $\overline{\alpha} = 3$.

Variable	Conditional probability vector
X_{s}	$P(X_{\mathrm{e}} = (h, l) \mid u_0) = (0.32{,}0.68)$
X_{s}	$P(X_{\mathrm{e}} = (h, l) \mid u_1) = (0.84{,}0.16)$
	$P(X_{\mathrm{s}} = (h, l) \mid u_2) = (0.44{,}0.56)$
X_{l}	$P(X_{\mathrm{l}} = (h, m, l) \mid u_3) = (0.20{,}0.34{,}0.46)$
	$P(X_{\mathrm{l}} = (h, m, l) \mid u_4) = (0.47{,}0.31{,}0.22)$
X_{h}	$P(X_{\mathrm{h}} = (n, y) \mid u_5) = (0.91{,}0.09)$
	$P(X_{\mathrm{h}} = (n, y) \mid u_6) = (0.82{,}0.18)$
	$P(X_{\mathrm{h}} = (n, y) \mid u_7) = (0.74{,}0.26)$

score of $Q(\boldsymbol{M}_{\text{CEG E}}) = Q(\boldsymbol{M}_{\text{CEG E}}) = -2,478.61$. This corresponds to a Bayes Factor of $\text{BF}(\boldsymbol{M}_{\text{CEG E}}, \boldsymbol{M}_{\text{CEG B}}) = 1.13$ very slightly in favour of the above analysed CEG models over B or D. So this would not usually be considered significant enough to prefer one model over another.

However, we would argue that CEGs B or D should be preferred to model the CHDS for three main reasons. First, CEGs E and F represent the same conditional independence structures associated with the childhood hospitalisation as CEGs B or D do but do so by using a much more complex graph. This makes these conditional independences somewhat less transparent and hence less intuitive for decision makers and domain experts. Second, there is no domain support for the variable orderings $\boldsymbol{X}(I_2)$ and $\boldsymbol{X}(I_4)$. Third, if the explanation of life events in terms of the family socio-economic status is a secondary objective then CEGs E and F are not able to provide us with this.

Finally, we can use the DP algorithm, Algorithm 5, to look over the CEG model space. Under the restriction of the assumed block ordering $\boldsymbol{\mathcal{B}}$ there are now two statistically equivalent CEGs given by the variable orderings $\boldsymbol{X}(I_1)$ and $\boldsymbol{X}(I_3)$. In particular, the CEG B given in Figure 7.5b is one of these MAP CEGs whilst the other one corresponds to a permutation between the variables X_{s} and X_{e} analogous to the one discussed for CEGs B and D above. As these models have already been studied in this section we will now proceed to analyse the CHDS dataset when no partial or total variable ordering is known.

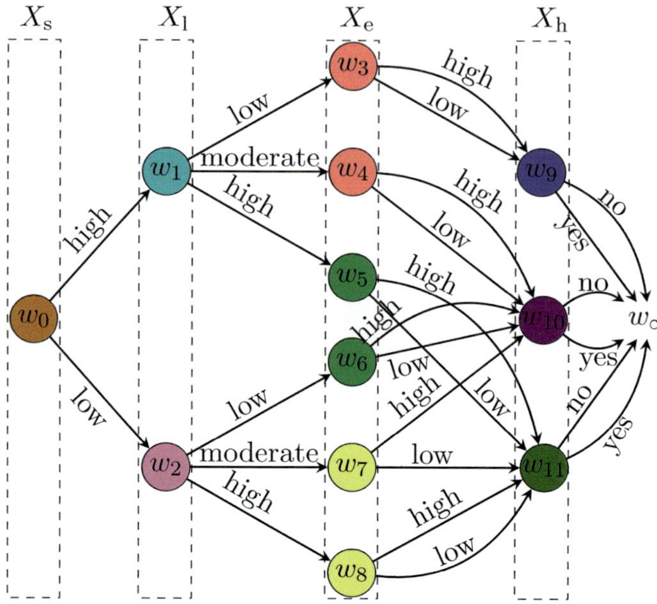

(a) The **CEG E** with variable order $\boldsymbol{X}(I_2) = (X_s, X_l, X_e, X_h)$.

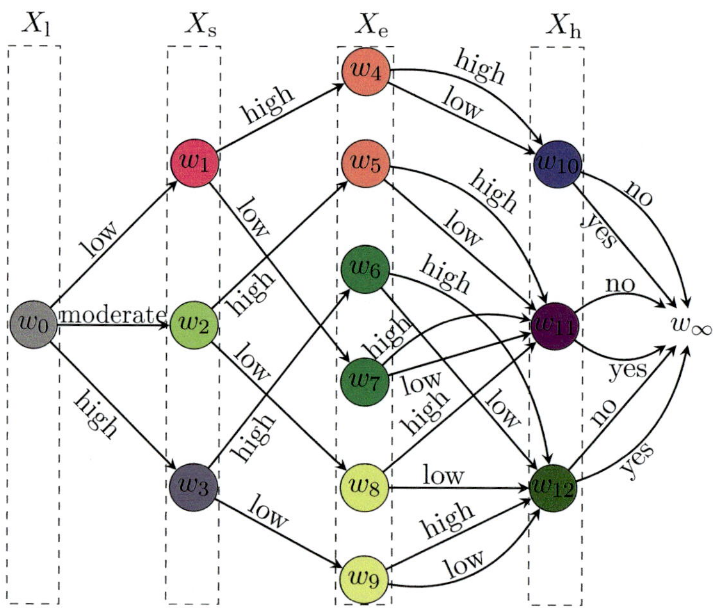

(b) The **CEG F** with variable order $\boldsymbol{X}(I_4) = (X_l, X_s, X_e, X_h)$.

FIGURE 7.9
CEGs found by the AHC algorithm using the CHDS dataset when we fix the variable orderings $\boldsymbol{X}(I_2)$ and $\boldsymbol{X}(I_4)$, respectively, and when we set $\overline{\alpha} = 3$.

7.4 Searching the CHDS dataset without a variable ordering

To truly explore the full CEG model space $\boldsymbol{C}(\boldsymbol{\mathcal{X}})$ of all possible explanations of the data which are compatible with the given problem variables, we could find the highest scoring CEG associated with each compatible event tree and then compare the corresponding scores. In the case of $\boldsymbol{\mathcal{X}}$-compatible event trees, this is implicitly and so more efficiently done by the DP Algorithm, Algorithm 5. In this type of approach we do not assume any prior information about the variable ordering for the CHDS dataset. Figure 7.10 depicts the two statistically equivalent optimal CEGs G and H resulting from such an analysis with conditional probability tables given in Table 7.7. The models $\boldsymbol{M}_{\text{CEG G}} = \boldsymbol{M}_{\text{CEG H}}$ correspond to the alternative optimal block ordering $\boldsymbol{\mathcal{B}}' = \{\boldsymbol{\mathcal{B}}'_1, \boldsymbol{\mathcal{B}}'_2, \boldsymbol{\mathcal{B}}'_3\}$ given by $\boldsymbol{\mathcal{B}}'_1 = \{X_{\text{s}}, X_{\text{e}}\}$, $\boldsymbol{\mathcal{B}}'_2 = \{X_{\text{h}}\}$ and $\boldsymbol{\mathcal{B}}'_3 = \{X_{\text{l}}\}$ since the variables X_{s} and X_{e} are exchangeable as discussed previously.

The score of these MAP CEGs is $Q(\boldsymbol{M}_{\text{CEG G}}) = -2,478.01$, corresponding to a Bayes Factor of $\text{BF}(\boldsymbol{M}_{\text{CEG G}}, \boldsymbol{M}_{\text{CEG B}}) = 1.57$ and $\text{BF}(\boldsymbol{M}_{\text{CEG G}}, \boldsymbol{M}_{\text{CEG C}}) = 1.14$ in their favour when compared, respectively, to the CEGs B and C from the previous section which are associated with the different variable ordering $\boldsymbol{X}(I_1) = (X_{\text{s}}, X_{\text{e}}, X_{\text{l}}, X_{\text{h}})$. These results would not usually be interpreted as giving significant evidence in favour of the block ordering $\boldsymbol{\mathcal{B}}'$ instead of the variable ordering $\boldsymbol{X}(I_1)$ or block ordering $\boldsymbol{\mathcal{B}}$ from the previous section. In fact, this may suggest that there is a dynamic interaction over time between the risk of childhood hospitalisation and the level of stress faced by their families where the family socio-economic status constitutes an explanatory background. For example, domain experts can hypothesise that not only family adversities increase the risk of hospital admission but also the hospitalisation of a child may put psychological and financial pressures over its family. Of course, this analysis goes beyond the objectives of this chapter but it does highlight how the CEG framework can provide decision makers and domain experts with useful insights. In this sense, it can stimulate some important alternative interpretations about the same phenomenon. We will discuss implications and putative causal interpretations of these types of observations in Chapter 8.

The block ordering $\boldsymbol{\mathcal{B}}'$ provides less analytical strength to study the hospitalisation rate than the variable ordering $\boldsymbol{X}(I_1)$ in the sense that in $\boldsymbol{\mathcal{B}}'$ the variable X_{h} is not the final one. As a consequence, here we actually loose the granularity provided by X_{l} to explain X_{h}. However, we need to recognise that the variable X_{l} is more prone to bias and manipulation since it relies only on the mothers' subjective reports. This can create some difficulties to legally and politically justify some medical policy that uses the criterion of life events to define a target population.

Now the two new CEGs G and H under the block ordering $\boldsymbol{\mathcal{B}}$ split children

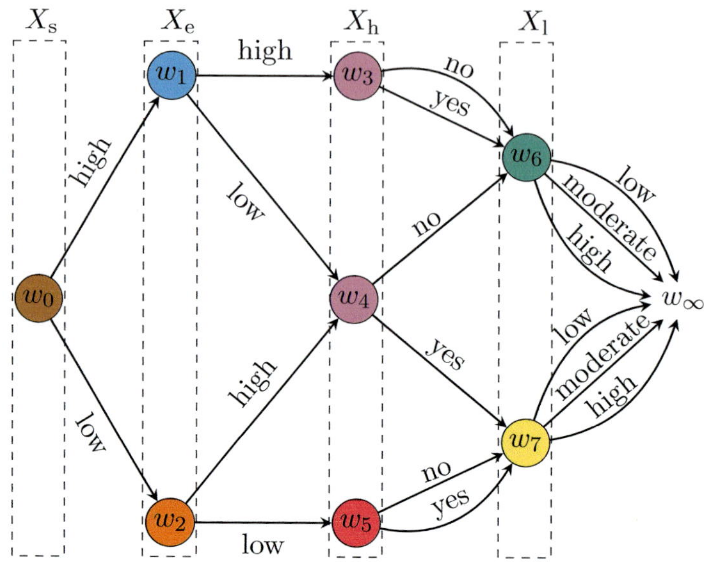

(a) The **CEG G** with variable order $\boldsymbol{X}(I_5) = (X_s, X_e, X_h, X_l)$.

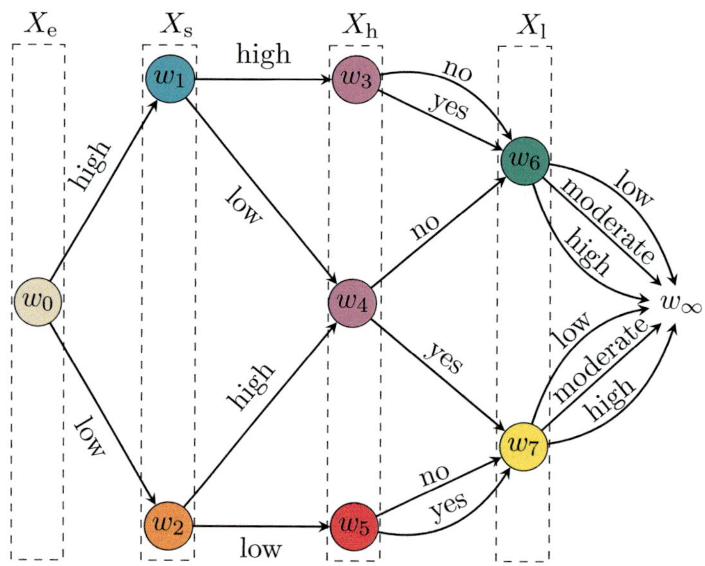

(b) The **CEG H** with variable order $\boldsymbol{X}(I_6) = (X_e, X_s, X_h, X_l)$.

FIGURE 7.10

Statistically equivalent MAP CEGs found by the DP algorithm using the CHDS dataset without restricting the variable ordering when we set $\overline{\alpha} = 3$. The stage structure of these graphs is given by: $u_0 = \{w_0\}$, $u_1 = \{w_1\}$, $u_2 = \{w_2\}$, $u_3 = \{w_3, w_4\}$, $u_4 = \{w_5\}$, $u_5 = \{w_6\}$, $u_6 = \{w_7\}$.

TABLE 7.7
Mean posterior conditional probability table for the **CEGs G and H** depicted in Figure 7.5 using the CHDS dataset when we fix $\overline{\alpha} = 3$. For CEG G, $Y = X_\mathrm{s}$ and $Z = X_\mathrm{e}$. For CEG H, $Y = X_\mathrm{e}$ and $Z = X_\mathrm{s}$.

Variable	Conditional probability vector	CEG G	CEG H
Y	$P(Y = (h, l) \mid u_0) =$	(0.57,0.43)	(0.32,0.68)
Z	$P(Z = (h, l) \mid u_1) =$	(0.47,0.53)	(0.84,0.16)
	$P(Z = (h, l) \mid u_2) =$	(0.12,0.88)	(0.44,0.56)
X_h	$P(X_\mathrm{h} = (n, y) \mid u_3) =$	(0.85,0.15)	(0.85,0.15)
	$P(X_\mathrm{h} = (n, y) \mid u_4) =$	(0.74,0.26)	(0.74,0.26)
X_l	$P(X_\mathrm{l} = (h, m, l) \mid u_5) =$	(0.18,0.34,0.48)	(0.18,0.34,0.48)
	$P(X_\mathrm{l} = (h, m, l) \mid u_6) =$	(0.46,0.31,0.23)	(0.46,0.31,0.23)

into two groups according to their risk of hospitalisation. Here, the higher risk group is constituted by children from socially and economically disadvantaged families. Therefore these models tend to highlight the most vulnerable children. This may be useful if policy makers need to focus on a very particular group using more objective and verifiable criteria because this separation is based only on a family's socio-economic status.

In contrast, CEG C which was the MAP $\boldsymbol{X}(I_1)$-compatible SCEG tends to clearly discriminate the group of children that have less chance of being hospitalised. This is because in the CEG C all children from socially and economically disadvantaged families are in the higher risk group which also includes children from other family background. Thus, the CEG C may be useful to design more inclusive policies when there is less financial pressure over policy makers. In this sense, splitting the children in three risk group of hospital admission, the $\boldsymbol{X}(I_1)$-compatible CEG B found by the AHC algorithm can provide decision makers with a compromise between these two approaches.

Also observe that the higher risk and the lower risk group of children have numerically very similar probabilities of hospital admission in each of the CEG models selected. However, the Hellinger distances between these risk groups in CEGs B, C and G (or H) are, respectively, 0.163, 0.122 and 0.097. So CEGs B and C provide us with a slightly better separation between these two groups of children in terms of numeric risk quantification than CEGs G and H.

Finally, with respect to life events, the CEGs B and C and the global MAP CEGs G and H identify two groups of families that have almost completely analogous probabilities of facing life adversities regardless of the model considered. The Hellinger distances between these two groups in CEGs B (or C) and G (or H) are slightly larger at, respectively, 0.225 and 0.235. However, the CEGs B and C include only socially and economically disadvantaged families in the higher risk group whilst CEGs G and H also add into this group families with divergent social and economic status whose children were hospitalised. So if decision makers aim at developing some social assistance policy models then B and C provide them with a more focused group of families whilst CEGs G and H enable them to design more inclusive policy that will probably be more financially demanding.

Overall we need to recognise that the choice of a 'best' model depends on the purpose of the application and all available domain information. Quantitative analysts should not recommend a model to support decision and policy makers and to domain experts based only on its Bayes Factor score but they should also consider the broad scope, understanding and impact of their studies. In this way, scientific reasoning and domain justification should not be replaced by a single statistical score, even a good one.

7.5 Issues associated with model selection

In this section, we illustrate some of the issues arising in using an exhaustive model search of the CHDS data and the necessity of using a heuristic algorithm to find a good SCEG model. These points were discussed in more generality at the very beginning of Chapter 6.

We conclude this chapter with a brief examination of CEG model selection using non-local priors and on how to set the hyperparameter $\overline{\alpha}$. The techniques we use in this development have been introduced in Section 6.4.

7.5.1 Exhaustive CEG model search

As has been outlined in Section 6.3, the computational complexity of the dynamic programming method for searching the SCEG model space is far higher than that for looking over its corresponding BN model space. This happens because the use of the DP algorithm (see Algorithm 5) for CEG learning requires us to consider all possible partitions of the situations at the many different levels of each event tree yielded by each permutation of the variable ordering—and we have shown that there are a Bell number of such event trees to look through. In particular, the computational complexity of finding the MAP CEG supported by a particular event tree is dominated by the number of vertices on the last non-leaf level of that tree.

For instance, the event tree for the CHDS data using the variable ordering $\boldsymbol{X}(I_1) = (X_{\text{s}}, X_{\text{e}}, X_{\text{l}}, X_{\text{h}})$ as in Figure 7.2 has 12 situations associated with its final variable X_{h}. This set of situations dominates the computational cost to search the model space spanned by $\boldsymbol{X}(I_1)$-compatible CEGs. In particular, here we will need to consider $B_{12} = 4,213,597$ partitions. This is very different from the variable ordering $\boldsymbol{X}(I_5) = (X_{\text{s}}, X_{\text{e}}, X_{\text{h}}, X_{\text{l}})$ which requires us to verify only $B_8 = 4,140$ partitions since the last non-leaf level has 8 situations associated with variable X_{h}. Thus the variable ordering $\boldsymbol{X}(I_1)$ yields approximately 1,018-times more partitions in the last non-leaf level than the variable ordering $\boldsymbol{X}(I_5)$. This is reflected in the system time necessary to find the MAP SCEG for each of these variable orders using the newly developed R-package `ceg` [82]. In a standard implementation carried out on a desktop with an Intel Core i7-4770 (3.40 GHZ) processor and 16 GB of RAM this takes 235.13 seconds for the ordering $\boldsymbol{X}(I_1)$ but only 0.23 second for the ordering $\boldsymbol{X}(I_5)$—a ratio of 1,022!

To understand this issue, suppose there was a single response variable Y added to the CHDS dataset and we had to explain it using the ordering $\boldsymbol{X}(I_1)$, depicting this variable as a new final level in an event tree. Then we would have to search over $B_{24} \approx 4.46 \times 10^{17}$ partitions yielded by the $12 \cdot 2 = 24$ situations associated with the variable Y. This is a factor of $^{B_{24}}/_{B_{12}} \approx 1.06 \times 10^{11}$-times the number of partitions compared to the ordering used in $\boldsymbol{X}(I_1)$. This scenario would get even worse if we had constructed social and economic variables with three different states rather than two. In that case, we would have to compute the score of $B_{27} \approx 5.46 \times 10^{20}$ partitions since the variable X_{h} would be represented by 27 situations in the corresponding event tree. This is a factor of $^{B_{27}}/_{B_{12}} \approx 1.30 \times 10^{14}$-times the number of partitions corresponding to the new variable X_{h} compared to the actual dataset with binary variables. In both cases using the processor and RAM above this would require us to spend more than *one billion years* to find the MAP SCEG given a specific variable order.

It is thus clear that it is only practical to use the exact dynamic programming algorithm for problems similar in size to the CHDS dataset. However, as the number of situations in the underlying event tree gets larger the method quickly becomes impractical and some approximate search algorithms need be employed.

7.5.2 Searching the CHDS dataset using NLPs

We have seen in the previous section that the use of efficient approximative algorithms for CEG model selection is necessary particularly when the complexity of the problem scales up. One appropriate option in this case is the use of the AHC algorithm (see Algorithm 2). When using Dirichlet local priors this algorithm may improperly merge situations into the same stages, particularly if these are situations not frequently visited by units in the sample. Using a non-local prior helps to reduce the incidence of such undesirable properties in the context of a greedy model search. These NLPs in particular enable us to find a sparser model that gives somewhat more transparent and intuitive explanations of the process and also has very good predictive capabilities.

To give a flavour of these very nice properties, we will now look over the CEG model space of the CHDS dataset using the AHC algorithm in conjunction with pairwise moment NLPS as introduced in Definition 6.5

in the previous chapter. We will hence again fix the variable ordering $\boldsymbol{X}(I_1) = (X_s, X_e, X_l, X_h)$ and also again set the hyperparameter $\overline{\alpha} = 3$. Regardless of the metric used to construct the pm-NLPs—Euclidean or Hellinger distances as in Theorem 4—the AHC algorithm now finds the same highest scored CEG. This CEG is henceforth denoted as CEG I and is depicted in Figure 7.11. Its statistically equivalent representations correspond to permuting the variables X_s and X_e. These would also have been found if we had adopted the ordering $\boldsymbol{X}(I_3) = (X_e, X_s, X_l, X_h)$. The conditional probabilities for CEG I are given in Table 7.8.

The CEG I can be interpreted as a compromise between the CEG B found by the AHC algorithm using Dirichlet LPs and the MAP CEG C: see Figure 7.5. All these CEGs exhibit the same conditional independence hypotheses for the first three explanatory variables. In line with CEG C, the CEG I indicates that there is evidence of only two hospitalisation risk groups of children: identified by the stages u_5 and u_6. However, these groups do not violate the assumption of probability dominance as CEG C does as discussed at the end of Section 7.2. In particular, CEG I *does not* identify children from socially or economically disadvantaged families living under moderate stress as having low risk of hospitalisation whilst keeping children from moderate stressed families that enjoy a comfortable economic and social situation as members of high risk group.

The higher risk group (in stage u_6) in CEG I contains the same children as the ones that are classified by merging the medium and high risk group of children (in stages u_6 and u_7) in CEG B. Recall from Section 6.4.2 that the main difference between the Bayes Factor scores provided by a Dirchlet LP and its corresponding pm-NLP is a penalisation term that the pm-NLP adds to the standard BF based on the distance between two stages that may be merged or not. Thus, an analysis of CEG I implies that although stages u_6 and u_7 in CEG B should be kept apart according to their probability mass, the distance between them does not provide sufficient support for such additional complexity in the model. In doing this, CEG I highlights only the substantial differences implied by the dataset and so provides us with new and much simplified hypotheses about how hospital admissions relate to the explanatory variables.

7.5.3 Setting a prior probability distribution

As discussed in Section 6.1, a key issue for CEG model selection is how to set the hyperparameter $\overline{\alpha}$. Following a recommendation for BN model selection [72] we have searched the CEG model space above using a setting of the hyperparameter that corresponds to the maximum number of categories taken by a variable in the CHDS dataset. This choice looks appropriate since it implies a plausibly large variance over the prior marginal distribution of each variable. However, some empirical studies have reported some sensibility of the BN model search to this setting of the hyperparameter: see e.g. [91, 93].

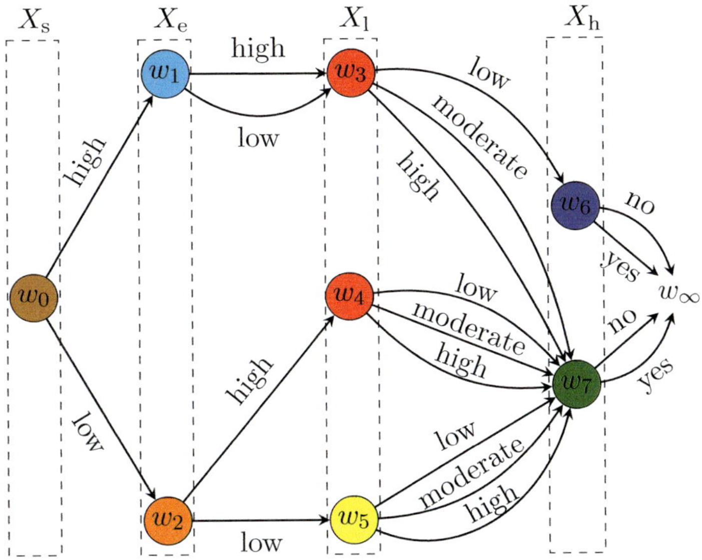

FIGURE 7.11
The **CEG I** found by the AHC algorithm in conjunction with Hellinger and Euclidean pm-NLPs using the CHDS dataset when we fix the variable ordering $\boldsymbol{X}(I_1) = (X_s, X_e, X_l, X_h)$ and set $\overline{\alpha} = 3$. The stage structure of this graph is given by: $u_0 = \{w_0\}$, $u_1 = \{w_1\}$, $u_2 = \{w_2\}$, $u_3 = \{w_3, w_4\}$, $u_4 = \{w_5\}$, $u_5 = \{w_6\}$, $u_6 = \{w_7\}$.

We have now performed an extensive computational analyses on the setting of the hyperparameter $\overline{\alpha}$ for CEG model selection based on the AHC algorithm in conjunction with Dirichlet local priors and pm-NLPs [17] and have briefly reviewed the practical implications of these results. Although the metric used to construct a pm-NLP might superficially look important, at least in the CHDS study the inferences appear to be robust to this choice. So here the recommendations for pm-NLPs are equally valid for Euclidean and Hellinger distances and, henceforth, we will not distinguish between these.

In particular, these simulation experiments were performed on the CHDS dataset assuming again the variable ordering $\boldsymbol{X}(I_1) = (X_s, X_e, X_l, X_h)$. The fixed generating model had a graphical structure corresponding to a slightly modified version of the MAP CEG C in order to eliminate the inconsistencies with regard to the probability dominance assumptions (see above). Its conditional probabilities were fixed based on the real dataset. For each sample size—ranging from 100 to $5,000$ by increments of 100—we simulated 100 complete datasets. Then for each dataset the AHC algorithm found the highest scored CEG model based on $\overline{\alpha}$-values ranging from 1 to 100 by increments of 1 and also on $\overline{\alpha}$-values of 0.1, 0.25, 0.5 and 0.75.

TABLE 7.8
Mean posterior conditional probability table for the **CEG I** depicted in Figure 7.11 using the CHDS dataset with $\overline{\alpha} = 3$.

Variable	Conditional probability vector
X_{s}	$P(X_{\text{s}} = (h, l) \mid u_0) = (0.57, 0.43)$
X_{e}	$P(X_{\text{e}} = (h, l) \mid u_1) = (0.47, 0.53)$
	$P(X_{\text{e}} = (h, l) \mid u_2) = (0.12, 0.88)$
X_{l}	$P(X_{\text{l}} = (h, m, l) \mid u_3) = (0.20, 0.34, 0.46)$
	$P(X_{\text{l}} = (h, m, l) \mid u_4) = (0.47, 0.31, 0.22)$
X_{h}	$P(X_{\text{h}} = (n, y) \mid u_5) = (0.91, 0.09)$
	$P(X_{\text{h}} = (n, y) \mid u_6) = (0.77, 0.23)$

Each CEG selected by the algorithm was then assessed according to two criteria.

1. *Total number of stages:* this captures the graphical complexity of a CEG.
2. *Total situational error:* this enables us to assess the predictive capability of the model. The error is calculated by the sum of situational errors over all situations in the event tree. The situational error corresponds to the Euclidean distance between the empirical mean conditional distribution and the generating conditional distribution of a particular situation in the event tree [17].

The final analyses were then conducted by averaging the values of each computed criterion over the 100 datasets for each pair of sample size and hyperparameter. Figures 7.12 and 7.13 shows our results for the sample sizes of 300, 900 and $3,000$.

Since the generating model had 7 stages, Figure 7.12 shows that the best results in terms of model complexity can indeed be obtained when the hyperparameter $\overline{\alpha}$ is not greater than 20. As expected, pm-NLPs tend to choose sparser CEGs and so CEGs which are closer to the true generating model than the Dirichlet LPs although the results for LPs are also good. In contrast to pm-NLPs, the AHC algorithm using Dirichlet CEGs appear not to improve substantially when the sample size increases.

In Figure 7.13 we can again see that $\overline{\alpha}$-values smaller than 20 provide better results, particularly for small sample sizes. Now all prior distributions select better CEG models when the sample size increases. The additional graphical complexity introduced by LPs does not deteriorate the predictive capabilities of the corresponding CEGs chosen when they are compared to those selected by pm-NLPs. This happens despite the pm-NLPs slightly dominating the LPs for a small sample size and when $\overline{\alpha}$-values are not large, and in medium and large sample sizes independently of the $\overline{\alpha}$-values.

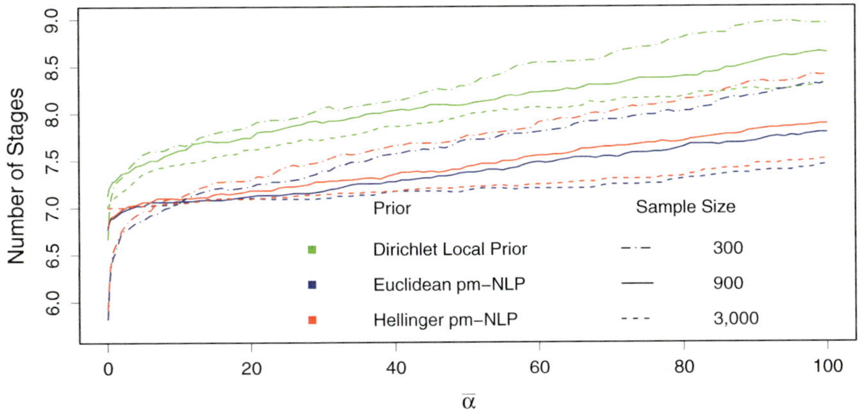

FIGURE 7.12
Average of the number of stages over the 100 CEGs chosen by the AHC algorithm in conjunction with Dirichlet LPs, Euclidean pm-NLPs and Hellinger pm-NLPs based on varying $\overline{\alpha}$-values for simulated CHDS datasets of sizes equal to 300, 900 and 3,000.

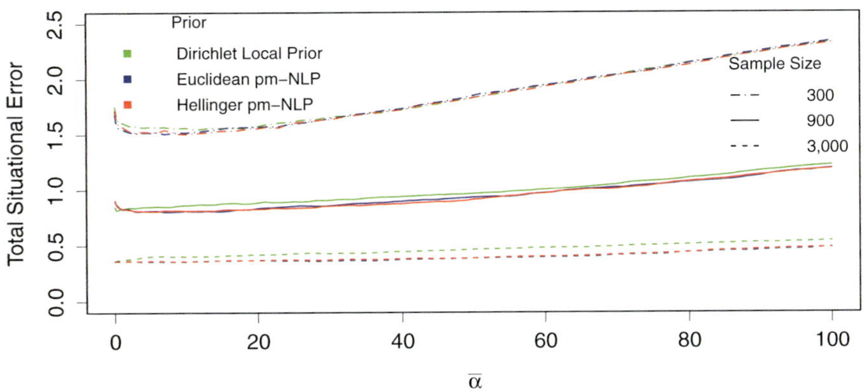

FIGURE 7.13
The average of the total situational errors over the 100 CEGs chosen by the AHC algorithm in conjunction with Dirichlet LPs, Euclidean pm-NLPs and Hellinger pm-NLPs based on varying $\overline{\alpha}$-values for simulated CHDS datasets of sizes equal to 300, 900 and 3,000.

Finally, we can analyse the impact of very small values of the hyperparameter $\overline{\alpha}$ of less than 1 on the CEG model search. These $\overline{\alpha}$-values yield results that are not reliable since the graphical complexity of the CEG model chosen by the AHC algorithm changes dramatically with small perturbation of $\overline{\alpha}$ in this range. These values are also prone to choose CEGs with larger total situation errors, especially for small sample sizes. Also note that the good modelling practice of setting a prior distribution often discourages the choice of such values.

Of course, large $\overline{\alpha}$-values are also not recommended unless we have genuinely strong information. In this study, such values are inclined to select more complex models that have larger total situational errors. Overall, the current recommendations for setting the hyperparameter $\overline{\alpha}$ in a BN framework look reasonable also in the CEG context. Note that the AHC algorithm using pm-NLPs tends to be more robust than one using standard Dirichlet LPs in the sense that CEG model selection is similar for wide intervals of values of the hyperparameter $\overline{\alpha}$ whose appropriate value is most difficult to prespecify faithfully.

7.6 Exercises

The following exercises refer to the artificial CHDS dataset available in the `R`-package `ceg`. Instruction on how to use the package can be found in [18].

Exercise 7.1. *Define a variable ordering on $X_{\mathrm{s}}, X_{\mathrm{e}}, X_{\mathrm{l}}, X_{\mathrm{h}}$ and plot the event tree depicting the state space of these random variables with that ordering.*

Exercise 7.2. *Assuming the variable ordering and the stage structure described in CEG I in Figure 7.11, plot that CEG and the corresponding staged tree.*

Exercise 7.3. *Suppose that you are interested in analysing how the covariates social background, economic background and life events may affect the response variable hospital admission. Using the optimised AHC algorithm implemented in the `R`-package `ceg` and choosing a Bayesian model selection approach with a hyperparameter $\overline{\alpha} = 3$, find the best scoring SCEG model for each of the six possible variable orderings. Plot the corresponding CEGs, analyse their posterior conditional probabilities and discuss the results. Will these results change if we set the hyperparameter $\overline{\alpha} = 10$?*

8

Causal inference using CEGs

The statistical study of causal relationships and causal inference has taken off over the last few decades and the literature on this topic is now vast. There is no way for this little book to do justice to this work or even to give a full review of either the theoretical development or the practical applications. However, what we can do is to indicate how the framework of the CEG currently adds to the armoury of this methodological toolkit.

Over the last thirty years there have been great advances in establishing and applying a sound mathematical framework for the rather philosophical notion of *causality*, or causal relationships that exist between a given set of problem variables: see for instance [66, 77, 90, 100] and references given therein. The main question researchers are concerned with in this context is how a change in one 'causal' variable influences changes in a collection of other 'effect|see causal effect' variables, where this change is measured relative to an observed system. This observed system we will also call the *idle* system. This question is not so much concerned with the nature of deterministic statements but rather with probabilistic relationships. Thus interest lies in whether there might be an effect felt, at least probabilistically, from one variable to another; and in the case when this effect is felt then what its strength might be, where strength is again quantified probabilistically in terms of an expectation or a probability mass function.

Whilst much of the early work on causality and causal inference started within the framework of counterfactual inference [87], a significant block of research has been spawned by the early work of graphical modellers and philosophers [77, 90, 100]. The latter took the standard BN—or its dynamic versions—and combined this framework with ideas taken from structural models to provide a sound methodology for enabling researchers to articulate two main points of interest. The first issue addressed was how potential causal relationships might be described to exist between a set of predefined problem variables. The second was then how when such a relationship was indeed assumed to be causal to measure the strength of the effect of one variable on another. The first of these developments provided the framework within which *causal discovery* could take place and the second defines what is sometimes called *causal algebra* or the theory of *causal manipulation* or *intervention.*

Causal inference has now been successfully applied to a vast range of problems all across science [10].

Because of the current state of the development of causation-based inference in CEGs rather than in BNs, we will focus most of our attention in this chapter on the problem of defining a sound causal algebra that can be used on the class of CEGs. This new framework, set up in Sections 8.2 and 8.3, will parallel Pearl's causal intervention calculus [77] as reviewed in Section 8.1. We will throughout use the BN notation first introduced in Section 2.3. Only at the end of this chapter in Section 8.4 will we give a brief review of recent advances in developing causal discovery algorithms for CEGs.

8.1 Bayesian networks and causation

We have seen that the CEG can directly through its graph embody collections of conditional independence relationships. It was precisely the analogous conditional independences within a BN—or equivalently the implicit factorisations of a joint density—that were used by Pearl and others [64, 66, 77] to formulate their causal algebras. These formally described what might happen if the modelled domain were to be subject to various types of control. Put simply, using expert judgements or a data-rich search, these authors first found the conditional independence relationships concerning an observed idle process expressible by a BN. They then extended this inferred qualitative structure to systematically construct numerous causal conjectures about what *might happen* were the process to be subject to various types of external manipulations or controls. To do this, formulae were defined which hypothesised that many of the conditional independences in the observed system would remain invariant under such interventions. Other features would be hypothesised to change—but in a predictable way. This gave a succinct way to *express* hypotheses about what might happen were someone to intervene in the system. But it also gave the apparatus to examine the implications of certain causal hypotheses about the *existence* of potential causal relationships. We will present the details of this development below.

Although the BN methodology has enabled us to study classes of very complex problems, its use for describing causal mechanisms that are embedded within discrete processes—especially where there is an intrinsic asymmetry in their structure—can have serious drawbacks. These problems were pointed out long ago by Shafer [90]. He proposed instead that causality could more generally and often more expressively be defined through the semantics of probability trees. Because Pearl's semantics based on a BN model are embedded in many of the discussions of causality, we need to explain in detail why there is a need for an embellishment of this technology so that it applies to CEGs. We do this below.

8.1.1 Extending a BN to a causal BN

Perhaps the simplest and most elegant method for extending a BN into a causal structure was given by Pearl [77]. He began by assuming that a BN describing the relationships that exist within a given population of units measured by a set of random variables $X_1, \ldots X_m$ had already been elicited. This could perhaps be drawn from a data analysis using a form of model search similar to the ones illustrated in Section 2.3 and Chapter 6. Suppose this BN has an associated probability mass function that factorises as

$$p(\boldsymbol{x}) = \prod_{i=1}^{m} p(x_i \,|\boldsymbol{x}_{\text{pa}(i)}) \tag{8.1}$$

for all states $\boldsymbol{x} \in \mathbb{X}$ as in (2.28).

Then Pearl argued that if that BN indeed fully and appropriately described the data-generating process and the BN truly described a 'causal' mechanism then this would lead us to assert the following were true. If we were to control the system so that the random variable X_j was forced to take the value $\hat{x}_j$ then this *atomic manipulation* of the BN would lead to a predictable consequence. The act of externally forcing a random variable to take a certain value is called a *do-operation* and usually denoted as $\text{do}(X_j = \hat{x}_j)$ for some state $\hat{x}_j \in \mathbb{X}_j$.

Under the notation of Section 2.3 and [64], Pearl asserts that if the underlying BN was truly causal then after such an atomic manipulation the joint probability mass function of the remaining variables in the system would be given by the formula

$$p(\boldsymbol{x}_{-j} \mid\mid \hat{x}_j) = \begin{cases} \dfrac{p(x_1, \ldots, x_j, \ldots, x_m)}{p(x_j \mid \boldsymbol{x}_{\text{pa}(j)})} & \text{if } x_j = \hat{x}_j \\ 0 & \text{otherwise} \end{cases} \tag{8.2}$$

where we use the symbol $||$ rather than $|$ to dinstinguish a probability mass function after intervention from a probability mass function after conditioning. Here, $\boldsymbol{x}_{-j} = (x_1, \ldots x_{j-1}, x_{j+1}, \ldots x_n)$ denotes the vector $\boldsymbol{x} \in \mathbb{X}$ with the jth entry deleted and $\boldsymbol{x}_{\text{pa}(j)}$ denotes the vector of values the parents of X_j take in a corresponding DAG representation, $j \in \{1, \ldots, n\}$. The right hand side of the formula above is also called the *effect* of the atomic manipulation $\text{do}(X_j = \hat{x}_j)$.

Suppose it were decided that an appropriate manipulation to consider would enforce a *collection* of $r \leq m$ different variables X_j to take particular values $\hat{x}_j$, where $j \in R \subseteq \{1, \ldots, m\}$ and $\#R = r$. Pearl proposes that the effect on this non-atomic manipulation should be a composition of its atomic effects. So we should apply the formula above r times. This would lead us to conclude that the probability mass function after this *composite manipulation* should be given by

$$p(\boldsymbol{x}_{-R} \mid\mid \hat{\boldsymbol{x}}_R) = \begin{cases} \dfrac{p(x_1, \ldots, x_m)}{\prod_{j \in R} p(x_j \mid \boldsymbol{x}_{\text{pa}(j)})} & \text{if } \boldsymbol{x}_R = \hat{\boldsymbol{x}}_R \\ 0 & \text{otherwise} \end{cases} \tag{8.3}$$

in direct analogy to the atomic intervention (8.2).

Definition 8.1 (Causal BN). *A BN is called* causal *if (8.2) and (8.3) are true for any type of manipulation of the system where no variable is forced to take more than one different value.*

At a first glance, the formulae (8.2) and (8.3) look a little mysterious. However, their interpretation is actually quite natural. In particular, (8.2) simply asserts that the effect on all variables X_i downstream from X_j—so all having a directed path in the DAG representation of the BN from X_j to X_i—will be predicted to be affected in exactly the same way as they would have been had we simply observed that X_j had taken the value $\hat{x}_j$ naturally. At the same time, the joint distribution of all variables not downstream of X_j would remain unaffected by the manipulation. Pearl would argue that if the BN were truly causal then we would naturally assume events associated with variables upstream of X_j to be unaffected by the manipulation because they would have occurred *before* those associated with the measurements of X_j. So their values already having been determined, the margin over these variables will remain identical to the one where X_j is not controlled in any way, $j \in \{1, \ldots, m\}$.

Of course assuming that (8.2) and (8.3) are true in applications is a very strong assumption. For example, in gene regulatory mechanisms, often the effect of a control whilst disturbing the upstream variables does not give rise to the same distribution we would get by conditioning: see for instance [69].

In any application of the formulae above we also need to be explicit about *how* we might plan to perform this control in a way that is as close to nature as possible. Only then could we check whether or not it might be a plausible assumption to make that the BN is causal in this sense within any given domain.

Despite these caveats such a causal assumption appears to be quite a natural one in a wide range of applications: particularly in scenarios where the describing variables reflect direct measurements of actual natural events and where the BN is fitted consistently with the order in which these natural events might happen. Because of this, the above approach to measuring causal effects has now been used very successfully in a range of applications performing a range of controls [10].

One important construction within all this technology is a measurement of the effect of a manipulation under the do-operator.

Definition 8.2 (Total effect). *Let $\boldsymbol{X} = (X_1, \ldots, X_m)$ be a random vector, let $j \in \{1, \ldots, m\}$ and let $Y(\boldsymbol{X})$ be any random variable measurable with respect to the probability mass function of $\boldsymbol{X}$. Then the* total effect *on Y of the manipulation* $\mathrm{do}(X_j = \hat{x}_j)$ *is the marginal probability mass function of $Y(\boldsymbol{X}_{-j})$ using the probability mass function of $\boldsymbol{X}_{-j}$ given in (8.2).*

In the definition above, Y is generally any function of the given problem variables. In applications this is often simply a projection onto a subset of the measurement variables. In this case we would be interested in the total effect of a control on a certain margin of the remaining variables.

There is now a significant literature determining when it is possible to estimate the total effect of a particular cause—so when an effect is *identifiable*—from an idle system when only certain margins of that system are observed [66, 85]. Analogues of this theory as it applies to CEGs are now available [104]: see also Section 8.3.3 below.

Following Definition 8.2, the *expected total effect* of the control $\text{do}(X_j = \hat{x}_j)$, for some $\hat{x}_j \in \mathbb{X}_j$, is simply a measurement of the predicted total effect on a variable Y of interest:

$$\mathbb{E}[Y(\boldsymbol{X}_{-j})] = \sum_{\boldsymbol{x}_{-j} \in \mathbb{X}_{-j}} y(\boldsymbol{x}_{-j}) p(\boldsymbol{x}_{-j} \mid\mid \hat{x}_j). \tag{8.4}$$

This is precisely the expectation of $Y(\boldsymbol{X}_{-j})$ with respect to the new mass function (8.2) hypothesised to hold after the atomic manipulation. In particular, if we find that

$$\mathbb{E}[Y(\boldsymbol{X}_{-j})] = \sum_{\boldsymbol{x}_{-j} \in \mathbb{X}_{-j}} y(\boldsymbol{x}_{-j}) p(\boldsymbol{x}_{-j} \mid\mid \hat{x}_j) = \sum_{\boldsymbol{x} \in \mathbb{X}} y(\boldsymbol{x}) p(\boldsymbol{x}) = \mathbb{E}[Y(\boldsymbol{X})] \tag{8.5}$$

is true then we might say that the manipulation of X_j to $\hat{x}_j$ has no measurable expected causal effect on Y. Suppose for instance that X_j was the weight of a patient and Y her blood pressure. Then were equation (8.5) to hold, the expected effect of a policy to control the weight of a typical patient, drawn at random from the population over which the BN applied, would be zero irrespective of the value her weight was controlled to.

We will henceforth think of a manipulated variable X_j as a *(putative) cause* of the downstream *effect* variables X_i which are involved in the calculation of the total effect of an atomic manipulation $\text{do}(X_j = \hat{x}_j)$ onto the remaining non-parent variables X_i in the system, $i \in \{1, \ldots, m\} \backslash \text{pa}(j)$. A more thorough introduction to the idea of putative causes is given by Pearl [77].

8.1.2 Problems of describing causal hypotheses using a BN

Despite their obvious successes there are three important issues which in our view prompt certain caution when defining causal hypotheses using the formula (8.2) and Definition 8.1. These make these BN methodologies less widely applicable than they might otherwise be. We separately analyse these points in this section.

Causes must happen before an effect

Many of the properties we might demand a causal relationship to exhibit are contentious. But one of the most universal and compelling demands we might make is that for some event to be attributed as a cause of an effect, that cause should *happen before* that effect. Almost synonymously we might reasonably want to assert that it must be possible to conceive of the effect as *a consequence* of that cause.

To illustrate why this is so consider the following example. Suppose a claimant in court asserts that her exposure to radiation from a particular nuclear accident caused her to develop a cancerous tumour. Suppose that, subsequent to her making this claim, incontrovertible evidence then came to light that the tumour existed within her body before the nuclear accident had occurred. In all courts of law this evidence would surely then be seen to demonstrate that the nuclear incident had not 'caused' the tumour to exist. Indeed to then designate the nuclear accident as a potentially contributing cause of the onset of the tumour would be seen as absurd.

This type of precedence requirement for an event (or variable) even to be a potential candidate as a cause of a different event must be almost universal. Outside perhaps examples drawn from abstruse areas of physics, any semantic associated with cause should have at its very core an expression of this idea.

Now we have already seen that the usual recommended way to construct a CEG is to draw its tree consistently with the way a domain expert perceives events unfolding: compare Section 4.1. Therefore within the very graph of a CEG can be embedded hypotheses about the order in which things might happen. This extra structural information is explicit in the representation and can be immediately harnessed to the service of the expression of causal hypotheses that respect the property discussed above. In particular, if a process is defined by a CEG, we can demand that causes can only be conceived to have an effect on an event if the cause corresponds to a set of situations closer to the root of that CEG and the effect sets of situation nearer the sink. This precedence condition can be read directly from the graph. Indeed, we saw in Section 4.2 that it is precisely because we need to provide a total order within the unfolding of events when specifying a CEG, and the invariance of the distributions within an observed system to some changes in that total order, which makes the identification of the statistical equivalence classes of CEGs so much more challenging than for a BN. However for causal inference this prespecified order is an asset.

In contrast, the BN automatically omits within its semantics some information about the order in which things happen even if this order is known to the modeller. At best the BN can only in general embody a *partial* order in the DAG representation. Of course some of the most experienced applied BN modellers [62] encourage us to ensure that when eliciting a BN the introduction of this partial order of variables be made consistent with any information about the order that exists about the event measured in the system but this is not formally required when specifying a BN. And indeed any extension of a BN to a causal BN as described above would make little sense unless this protocol were followed.

To see that precedence ordering can be suppressed even in the simplest of

BNs, suppose we may know that one random variable X is a measurement of an event which is known always to happen before the consequences measured by a second random variable Y happen. Suppose X is independent of Y. Then the DAG representing this system has vertices $\{X, Y\}$ and no edge. So whether what happens as measured by X does happen before or after the events measured by Y is not represented by the BN model.

Before we take observations that convinced us of the independence of X and Y we might have plausibly entertained the possibility that X might cause Y. Only after collecting data might we conclude on the basis of this evidence, for example using the adage 'no causation without correlation' that there appears to be *no significant measurable causal effect* of X on Y. So in this case an atomic manipulation of X might have zero expected total effect on Y. But that is very different from the assertion that Y could *not logically* have ever been thought of as a cause of X. So there is an asymmetry here in the causal role of X and Y which cannot be captured by the BN. As a BN becomes more and more complex these sorts of ambiguities become more and more nuanced. In contrast, by explicitly expressing a hypothesis about the order in which things can happen, the CEG framework provides a much less hazardous and more complete framework in this regard.

Causes are events not random variables.

The current ways of embedding causal hypotheses within a BN and the development of causal discovery algorithms reflects the way data is most often collected. This can for instance be in form of a matrix which has rows labelled by the observed units in a sample and the columns labelled by a type of measurement taken on each of these units. We are then drawn into thinking of that observed information in terms of replicates of a random vector of observed measurements within a given population. Causal inference can hence be expressed in terms of the conditional independence structures that seem to appear across the components of those vectors.

However, we would contend that at a primitive level, probabilistic causal relationships need to be expressed as relationships within the probability space itself rather than as a set of measurements on that space. For example, suppose interest lies in a person's weight. Someone notices that a set of weighing scales—our measuring instrument—measures 3lbs more in the bedroom than in the bathroom. By treating measurements of weight and weight as synonyms we would be able to assert that moving the scales into the bathroom would cause that person to lose weight. When any standard concept of weight loss is the effect of interest then this 'causal' deduction would not be very apposite.

There is a further related but even more profound issue associated with causal statements. Consider the statement that an earthquake of magnitude 8.0 or above, more than 1km beneath the sea, might cause (in a probabilistic sense) a tsunami. When you reflected on the veracity of this statement do you think only of the possible effect that a tsunami is generated by such an earthquake, or did you find it necessary also to think about what might happen were there no such earthquake? We would contend that causal statements like this have a causal meaning irrespective of what else might have happened.

To express this statement using hypotheses about random variables requires much more architecture to be added. In particular, the tree of possibilities needs to be dramatically extended to describe events which were not implied by the earthquake. In order to construct any *random variable* X to be a cause, we would first need to introduce other levels it might take and to build a probability model to describe these other circumstances. So within a BN framework we also need to set up and quantify what might happen were X to take other levels that together gave an exhaustive coverage of all possible other things that might have happened (for example, earthquakes of different severity, volcanic eruptions, a nuclear test, something other than all of these articulated possibilities). Do we really need to do this to consider the statement above? We think not.

Thus whilst the expectation of a particular random variable might often be a natural expression of the expected effect of a cause, it is much less natural to think of a cause as a random variable. Even within the BN framework, a causal intervention can be associated with a decision to control a variable to take a *set value* as in the do-operation $\mathrm{do}(X = \hat{x})$ we analysed in the previous section. But a random variable taking that value corresponds to the *event* $X = \hat{x}$ in the set of possible outcomes. So why not think of causes simply in terms of such events—which in any case we would argue are the objects of real interest?

Context specificity of acts and when BNs are not faithful

The above two issues are foundational ones that might lead us to examine the advantages of tree-based methods—like using CEGs—in preference to those based on BNs. There are two rather more pragmatic but nevertheless important issues that favour working with CEGs. These both concern the nature of the dependence structure within the observed (uncontrolled) system and so link to points we have made in the previous chapters of this book. These remain important in this causal setting and indeed have special causal implications.

The first issue is that often even in the uncontrolled process it is not natural or efficient to describe the dependences through a BN even if this is formally possible. This is especially true when the underlying tree of the process is very asymmetric: we have seen many examples of this in Chapter 3. If we were to describe the process symmetrically then we would need to first construct dummy states which are then assigned a zero probability of occurring. In causal inference, such an extension although formally sound for describing an observed system can lead to absurd interpretational labelling. Extending causal hypotheses to these already artificial categories has the potential to create great confusion.

Second we have seen that conditional independence results within a CEG apply to specific random variables that are deduced from the graph. In contrast it has been discovered in practice that even when a BN is valid it can contain many context-specific independences. This typically means that a condition called *faithfulness* needed for most causal discovery algorithms is violated: see also Section 2.3.

Re-expressing context-specific conditional independence relationships in a BN within the stage structure of a CEG has then two advantages. First, we can often articulate sets of possible cause and effect variables to later investigate directly within our coloured graph. Second, because the CEG class can express a much richer set of conditional independences than a BN, the analogue of the faithfulness condition used for BN discovery—whilst still constraining—is much less severe in the CEG than it is for a BN. In particular, all context-specific BNs remain faithful in their CEG representation [98].

8.2 Defining a do-operation for CEGs

The analogy of the do-operation for BNs (8.2) as it applies to a probability tree turns out to be extremely straightforward and indeed much more transparent than for the BN. In (staged) trees and CEGs, we no longer think of a manipulation as applying to a random variable. We instead think of this manipulation as forcing any unit that arrives at one of the situations of an event tree to proceed along one of the edges emanating from that vertex. A manipulation in a BN leaves untouched what has happened before the manipulation whilst exciting the same response on individuals as what might have happened had the result of the manipulation occurred naturally within the idle system.

Following exactly the same interpretation of an intervention immediately gives us the following definition, using notation from Section 3.1.

Definition 8.3 (Tree-atomic manipulation). *Let $(\mathcal{T}, \boldsymbol{\theta}_\mathcal{T})$ be a probability tree. Let $\mathcal{T} = (V, E)$ be the graph of that tree and let $\hat{e} = (v, v') \in E$ be an edge within this graph. The effect of the* tree-atomic manipulation *of forcing any unit arriving at the situation v to proceed along the edge $\hat{e}$ produces a new (degenerate) probability tree $(\mathcal{T}, \boldsymbol{\theta}_\mathcal{T})_{\hat{e}}$. This new tree has the same graph $\mathcal{T}_{\hat{e}} = \mathcal{T}$ but the edge probability of the enforced edge is set to one, $\theta(\hat{e}) = 1$, and the probabilities of all other edges $e' \in E(v) \setminus \{\hat{e}\}$ emanating from v are set to zero, $\theta(e') = 0$. Otherwise, the new tree inherits all edge probabilities from $(\mathcal{T}, \boldsymbol{\theta}_\mathcal{T})$.*

Every probability tree is a graphical representation of an underlying set of atoms Ω. So a tree-atomic manipulation changes the atomic probabilities $p_{\boldsymbol{\theta}}(\omega)$ of every atom $\omega \in \Omega$ in that space which are associated with a path

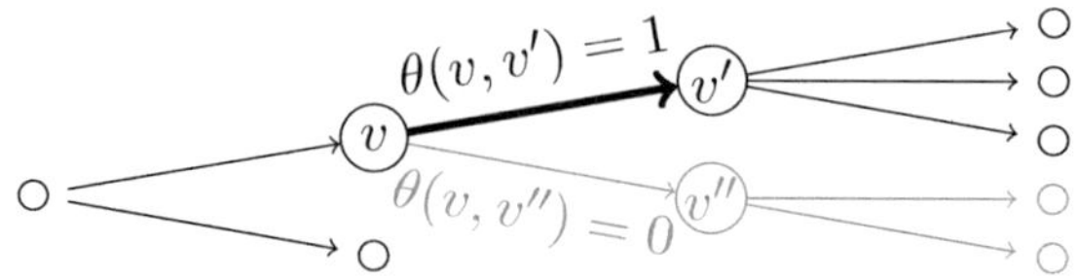

FIGURE 8.1
An intervention, or tree-atomic manipulation, on the edge $\hat{e} = (v, v')$ in a probability tree.

containing the edge $\hat{e}$. The new probability of such an event is

$$p_{\hat{e}}(\omega \mid\mid \hat{e}) = \frac{p_{\boldsymbol{\theta}}(\omega)}{\theta(\hat{e})} \tag{8.6}$$

and equals zero for all atoms ω associated with root-to-leaf paths passing through v but not through the edge $\hat{e}$. All atomic probabilities which are represented by root-to-leaf paths not in $\Lambda(v)$ are thus not affected by this manipulation. Consider Figure 8.1 for an illustration.

Because the manipulated probability tree includes edges with labels taking the values zero and one, it lies on the boundary of our initial definition of a probability tree and is what we have called a *degenerate* tree in Sections 3.1 and 3.3. If this probability tree is a staged tree then its stage structure can of course be violated by a manipulation which changes the values on only one floret in a stage. The manipulated vertex will then not be in the same stage as the remaining vertices. As a consequence, the CEG corresponding to the idle tree can then be very different from the CEG corresponding to the manipulated tree: we explore this observation in Exercises 8.1 and 8.2 given below. In particular, the CEG associated to a manipulated staged tree is also degenerate.

The new formula (8.6) is a direct analogue of the BN formula (8.2) above. It simply transforms a probability mass function p to a new mass function

$$p_{\hat{e}} = g_{\hat{e}}(p) \tag{8.7}$$

using a map $g_{\hat{e}} : p \mapsto p_{\hat{e}}$ which enacts the tree-atomic manipulation given in Definition 8.3.

Importantly, this new density can be interpreted as a density over a different (non-degenerate) probability or staged tree whose graph we denote $\mathcal{T}_{-\hat{e}}$ for $\hat{e} = (v, v')$. This new tree arises from the manipulated tree $\mathcal{T}_{\hat{e}}$ by erasing all root-to-leaf paths associated with unfoldings of situations made impossible by the manipulation. These will be all subtrees $\mathcal{T}(v'')$ rooted after edges $e = (v, v'')$ that had their attached probabilities set to zero, so all edges with $v'' \neq v'$ will disappear, and the enforced edge $\hat{e} = (v, v')$ is erased, so now $v = v'$. We assume all labels from the original tree that are not associated to erased edges to be inherited by the new tree.

In Figure 8.1, the new tree would arise from erasing all greyed-out edges and vertices, the thick depicted enforced edge (v, v') would disappear and the edge emanating from the root which does not end in v—and all its future unfoldings—would be unaffected.

Clearly, the labels of $\mathcal{T}_{-\hat{e}}$ now induce a corresponding probability mass function which is identical to the one of the manipulated tree, $p_{-\hat{e}} = p_{\hat{e}}$, but which has a graphical representation that is not degenerate. Naturally, the corresponding CEG $\mathcal{C}(\mathcal{T}_{-\hat{e}})$ will not be degenerate either.

As a consequence, the map $g_{\hat{e}}$ from (8.7) is a well defined map that takes one proper—non-degenerate—CEG to another proper CEG. In particular, the space of proper CEGs is *closed* under the atomic manipulation defined above. In addition and in contrast to the BN methodology developed in the previous section, tree-atomic manipulations are very easily enacted graphically in the sense that the graph of a staged tree or CEG can tell us directly which unfoldings disappear and which are not affected by a control. We develop an example in Section 8.2.2 which illustrates this point.

For obtaining a most general tree-atomic manipulation operation, it was necessary to develop Definition 8.3 in terms of probability trees (or staged trees) rather than CEGs. This is because when manipulating edge labels in a CEG, in the underlying tree we would implicitly not only change the labels of single edges but of all edges whose vertices are in the same position. So this is only a special case of the tree-based definition we provide here: see also the following section. However, we discuss in Section 8.3 that sometimes the CEG type of position manipulation is indeed appropriate and more transparent to a context of interest: compare Definition 8.4. In the next section we show that the do-operation in BNs is only a special case of such a CEG manipulation.

8.2.1 Composite manipulations

In practice, a decision maker will often choose or be constrained to manipulate many situations at the same time. The hypothesised causal effect of such composite manipulations has already been defined for a BN: see (8.3). It is sufficient here for us to simply copy the rationale for BNs and thus obtain a formula for the hypothesised effect of such a composite manipulation on a model whose observational process is described by a staged tree.

Suppose we plan to enact tree-atomic controls (8.7) on units arriving at a situation v_1 at the head of an edge $\hat{e}_1 = (v_1, v_1')$ forcing the units along that edge and enacting another control on the same tree, this time on units arriving at a different situation v_2 at the head of $\hat{e}_2 = (v_2, v_2')$ forcing these units along this other edge, $v_1 \neq v_2$. Then we simply define the hypothesised effect of these two manipulations on the mass function of the controlled system as the composition of the two effects:

$$p_{\hat{e}_1\hat{e}_2} = g_{\hat{e}_2}(p_{\hat{e}_1}) = g_{\hat{e}_2}(g_{\hat{e}_1}(p)) \tag{8.8}$$

where p denotes the probability mass function of the idle system. If in this setting v_2 is downstream of v_1 and in particular is downstream of an edge (v_1, v_1'') which has been assigned probability zero, so $v_1'' \neq v_1'$, then the second manipulation simply has no effect.

It is easily checked that (8.8) is a probability mass function on a new set of non-zero atoms provided that the manipulation is logically possible. In particular, this composite of two atomic manipulations on a non-degenerate CEG will only be impossible if the two situations are the same $v_1 = v_2$ but the enforced edges are not $\hat{e}_1 \neq \hat{e}_2$—so only if we simultaneously force units arriving at v_1 to both go down development $\hat{e}_1$ with probability 1 and also $\hat{e}_2$ with probability 1.

If instead we tried to use the formula above for this infeasible composition then the manipulated tree $\mathcal{T}_{\hat{e}_1\hat{e}_2}$ would formally be defined to have two emanating edges which are assigned probability 1. It would follow that on that particular floret, $\theta(\hat{e}_1) + \theta(\hat{e}_2) = 2$. So the illogicality of the proposed manipulation would give rise to a mass function p that is no longer a *probability* mass function on the atoms $\omega \in \Omega$. Henceforth a composition of two atomic manipulations is called *infeasible* if and only if the mass function described above is not a probability mass function.

The order in which we perform a feasible composite manipulation does not matter, in the sense that always

$$g_{\hat{e}_1\hat{e}_2} = g_{\hat{e}_2\hat{e}_1}. \tag{8.9}$$

A proof is provided in [46].

This result is important if we need to define a manipulation of a system where these controls are imposed simultaneously: whichever order we choose will give rise to the same hypothesised effect.

The definition above of course extends to the case where we might want to enact r separate atomic manipulations simultaneously. Thus suppose for a set of situations $\{v_1, v_2, \ldots, v_r\} \subseteq V$ we want to enact a control that whenever a unit arrives at a situation v_i we force it down a fixed emanating edge $\hat{e}_i = (v_i, v_i')$ with probability 1, for all $i = 1, 2, \ldots, r$. Then we simply define this composite control as

$$g_{\hat{e}_1\hat{e}_2\ldots\hat{e}_r} = g_{\hat{e}_r} \circ \cdots \circ g_{\hat{e}_2} \circ g_{\hat{e}_1}. \tag{8.10}$$

Again it is easy to check that this control is well defined and does not depend on the particular order we chose for the edges within the set above. The stage structure is hereby sequentially modified in the way we describe it above, resulting in a manipulated tree $\mathcal{T}_{-\hat{e}_1\hat{e}_2\ldots\hat{e}_r}$. As a consequence, the space of CEGs is closed under composite tree-atomic manipulations.

Importantly, whenever a stratified staged tree or SCEG is compatible with an underlying set of random variables which can alternatively be represented by a BN, then a composition as in (8.10) can be employed along the same level of the graph in order to enforce a manipulation directly analogous to Pearl's atomic intervention in (8.2) [104]. In fact, all interventions considered by Pearl can be represented within our framework and give the same formulae for the corresponding probability mass function. The development above is thus a direct generalisation of Pearl's causal algebra—expressing his development as a special case of ours.

8.2.2 Example: student housing situation

We will now analyse causal manipulations in an extension of the running example from [46] which was motivated by the real system used in [104].

Suppose we observe data on two groups of students: those who win a place at Oxford and those who win a place at Warwick University. Suppose our students are assigned to either of these groups with positive probability. Then to simplify this setting we assume that first year students at Warwick University can find accommodation either on campus, in Coventry or in nearby Leamington Spa. Landlords in both cities can be either friendly or grumpy. If they are grumpy, then students might move house, and if they do move then they might also leave the city they live in. We shall assume that there are no landlords on campus and that the attitude of landlords is the same in both Coventry and Leamington Spa, so the probability of renting with a friendly landlord does not depend on the location. We also assume that, within this subgroup, we are only interested in the flow of students living in Coventry and renting with a grumpy landlord. Similarly, students at Oxford University can in this simplified setting either rent private accommodation or live in a college. Those who are not assigned a room in college are subject to the same landlord situation as above, and we are again interested in the behaviour of those renting with grumpy landlords.

We can represent this setting by the staged tree in Figure 8.2 where the blue-coloured vertices v_3, v_4, v_5 are in the same stage because the attitude of landlords is assumed to be independent of their location.

As an alternative, we could model this example using a BN represented by the collider DAG in Figure 8.3. The vertices of this graph correspond to the discrete random variables measuring which university a student went to, so X_{u} taking values *Warwick* or *Oxford*, modelling the type of accommodation X_{a} taking values *Coventry*, *Leamington Spa*, *Oxford* and *College*, a variable for the attitude of the landlord X_{l} taking values *grumpy* and *friendly*, and measures of relocating within a city X_{r} and moving away X_{m}, the latter two taking values *yes* and *no*. Following the problem description above, X_{a} and X_{l} are independent, and X_{m} is independent of all other variables provided that X_{r} is known.

Comparing Figures 8.2 and 8.3, we see as with examples presented in Chapters 3 and 7 that again the staged tree is in this setting a much more transparent tool for communicating our asymmetric modelling assumptions.

Suppose we are now interested in what would happen were Warwick stu-

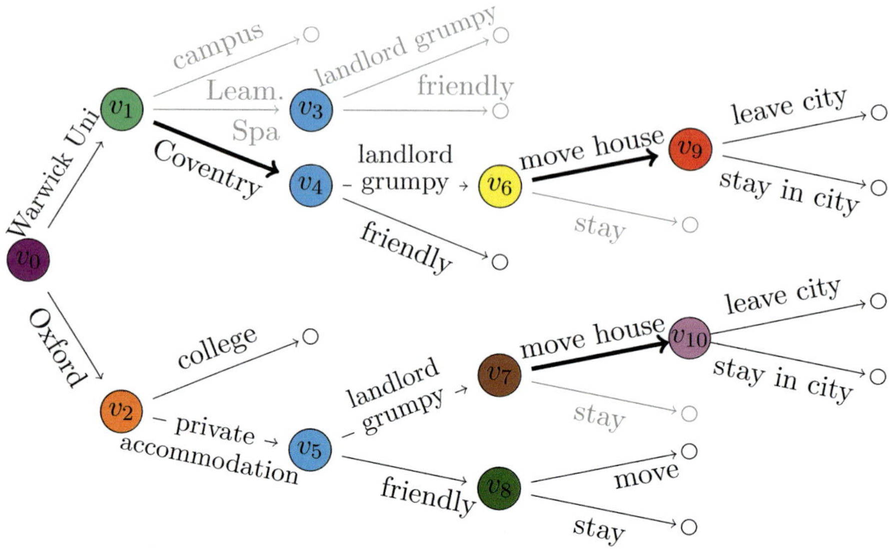

FIGURE 8.2
A staged tree representing the accommodation situation at two universities. Manipulations are performed on different branches of the tree.

dents forced to live in Coventry, for instance by a university policy which aims to cut time spent commuting to Leamington Spa and because accommodation on campus will be subject to refurbishment. We can model this policy by enforcing a tree-atomic control in the staged tree. Under this intervention, the edge labelled *Coventry* would be the only possible unfolding subsequent to the edge labelled *Warwick Uni*. We depict this by the thick black edge in Figure 8.2 and assign probability one to this edge. So the greyed-out subtrees in that figure disappear when passing from the idle to the manipulated system: these unfoldings are assigned probability zero. The resulting staged tree—and its corresponding CEG—are now degenerate. However they are statistically equivalent to a new staged tree where the greyed-out subtrees are not present and the edge with probability one is cancelled, merging the vertices ending in *Warwick* and emanating from the floret labelled *landlord* as in Figure 8.1 above.

In other words, when enforcing this control the atomic probabilities depicted in the staged tree are projected onto a subtree. As a result, under our causal hypothesis the new probabilities—calculated as in Definition 8.3—can again be simply read off the labels of the graph. For instance, the probability of a student who is renting with a grumpy landlord leaving the city of Coventry when she was initially forced to live there is simply the product along the relevant root-to-leaf path in the manipulated system. Compare also Exercise 8.3 below. In fact, this type of calculation can be elegantly performed

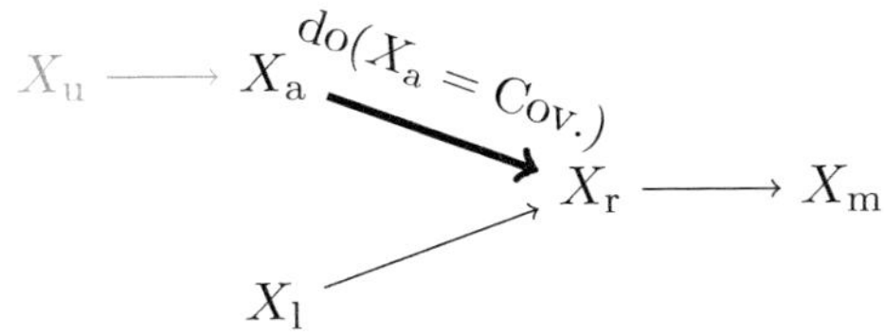

FIGURE 8.3
A DAG representation for the student accommodation example. Interventions are a lot less straightforward in this graph than in the alternative staged tree from Figure 8.2.

using a differential operation on the interpolating polynomial of this staged tree: see [43] for a full account of these calculations.

In the DAG representation of this example, the same control would force the accommodation variable to take the value Coventry, so do($X_a = Coventry$) in Pearl's notation. If the BN is causal then the effect of this intervention can be calculated using the semantics of the do-operator introduced in (8.2) above. However, the calculation of an effect as above cannot be read from the DAG in a straightforward manner. Instead, it needs to be elicited using the rules of do-calculus. By using a DAG a crucial point is also obscured: because do-interventions are independent of the ancestors of the point of intervention, the university variable X_u which was an ancestor in the idle system does now not influence any effects in the manipulated system. For example, those students who study at Oxford are clearly not affected by the policy we introduced at Warwick. As a consequence, for this subgroup the do-operation does not make sense—and the DAG has no semantics to take care of this subtlety. Thus, a 'contex-specific' type of intervention cannot be transparently performed in a BN using Pearl's original semantics.

Suppose now that in the framework of a mental health programme the Student Unions of both Oxford and Warwick University campaigned for students renting with a grumpy landlord to move house. We are interested in assessing the effect of this intervention on students: will they not only leave their accommodation but also leave the city they live in? In order to calculate this effect, we would now intervene a second time on the idle system in Figure 8.2, assigning probability one to all unfoldings labelled 'move' which appear after an edge labelled 'grumpy'. This is again depicted via thick edges and greyed-out alternative unfoldings. We can assess the desired effect in the same fashion as the effect above, with details presented in [43].

Both policies—those concerning the location of accommodation and also those concerning mental health—could have been performed in either order and would have in both cases provoked the same effects on the student population. This is the result we found in (8.8) above. As a consequence, the staged tree representation of this problem not only transparently depicts the asymmetric modelling assumptions and our vertex manipulations but always yields a straightforward framework for performing composite manipulations on very different points of intervention in any chronological order—of course with the rider that these are consistent with the problem. This is a huge advantage of the staged tree over a BN as a framework for expressing nuanced causal hypotheses. In particular, in more complex settings and for instance in applications in health studies, staged tree models can prevent absurd interventions such as formally assigning treatment to patients which have already died: a subtlety not transparent in DAG representations of BN models.

8.2.3 Some special manipulations of CEGs

There are two important types of controls on CEGs which can embed Pearl's ideas in a meaningful more general way. The first was originally introduced in [105] and is called a *positioned manipulation*. This manipulation forces all units arriving at a vertex in a certain position along a fixed edge, and simultaneously enforces this control for all vertices in that position along edges with the same meaning. These can be thought of as atomic manipulations of the edges of the CEG—rather than the edges of the corresponding staged tree as in the development in the previous section. This in fact sets the vertex-random variables from Section 4.1.4 in a CEG to a particular value, namely the label of the enforced edge: in symbols, $\mathrm{do}(X_w = w')$ for w and w' situations in a CEG and connected by an edge. This is a useful type of operation since the same control is always applied to units who exhibit identical prospective developments in the future. As a consequence, it links closely to any definition of causation which uses positions as primitives: see the discussion of Definition 8.4 below.

A second important subclass is the *staged manipulation*. This is analogous to the positioned manipulation above but this time we make the more stringent demand that if the control contains an edge in one situation in a certain stage it contains an associated edge in all situations in the same stage—rather than in the same position. All atomic manipulations considered by Pearl are in particular compositions of stage manipulations.

8.3 Causal CEGs

We have discussed above how a system represented by a staged tree or CEG can be given a useful causal extension, paralleling Pearl's development for BNs. As in Definition 8.1, Pearl originally called a BN a causal BN if the causal formula (8.2) remains valid for *all* values that the variables in the system might be manipulated to take. Even in the BN setting it has been subsequently found that this demand is an extremely stringent one and indeed often does not make sense. For example, if one of the measurement variables were the height of a person then even the concept of manipulating that height to a different value is problematic. Thus we typically would only want to demand Pearl's formula to hold for those variables it might be logically possible to manipulate and to demand that indeed only these values we will consider manipulating units to.

8.3.1 When a CEG can legitimately be called causal

Demanding that a CEG is causal to all possible manipulations is an order of magnitude more stringent than the demands of a general causal BN in Definition 8.1. So it is even more of a priority in this setting to prescribe the set of manipulations we are prepared to demand the causal formula (8.6) to hold for.

This can in fact be very simple done within the semantics above.

Definition 8.4 (Causal CEG). *Let A be a collection of situations in an idle staged tree. We call the corresponding CEG* A-situation/stage/position causal *if for all situations/stages/positions in all subsets of A the causal formula (8.6) under a control is believed to be valid.*

The notion above is in direct analogy to the notion of causality used by Pearl [77]. Assuming this condition holds for the entire CEG, so that A is the complete vertex set, is often unnecessarily strong [105]. Hence, the set A will usually be defined in relation to the purpose of the study. In practice, we hence often find that it will be quite small. See also Exercise 8.4.

For instance in our toy example in Section 8.2.2, all proposed interventions were interventions on situations. The intervention of forcing Warwick students to live in Coventry is not an intervention on a position. The intervention to force potentially unhappy students to move house is an intervention both on stages and on positions. We have assumed above that the causal formula holds for these positions: so the given staged tree (and its corresponding CEG) are situation or stage and position causal as in the definition above.

So we can now set up the analogy to Definition 8.2 in terms of CEGs.

Definition 8.5 (Causal effect). *The* total effect *of a manipulation of an idle CEG on an atom as in Definition 8.3 is given by the value of the probability mass function (8.6) of the manipulated CEG evaluated at that atom.*

Again in analogy to the definition for BNs, the expected total effect on a random variable is then simply the expectation of that variable under the new probability mass function.

8.3.2 Example: Manipulations of the CHDS

For a CEG the actual measurement of a total effect as in Definition 8.2 on a random variable given a causal manipulation is simple to calculate whenever we have a completely observed system. This is because we have shown in Section 5.1 how, on the basis of a complete dataset, we can estimate floret parameter vectors. Under the appropriate causal hypotheses, the distribution of an effect after a given causal manipulation is simply expressible as a function of these parameters—so the edge probabilities defining the CEG. When we have performed the usual conjugate analysis, the joint distribution of these parameters is simply a product of Dirichlet densities. Any total effect can then be calculated as an expectation of a given function of these posterior conditional probability vectors. So the predictive distribution can be calculated precisely. We illustrate this in an example below.

Of course when we allow for uncertainties in estimation of the idle system for BNs then such estimates can be uninformative: see for instance [86] for a discussion of a related issue. Even if the prior distributions are not chosen to be conjugate, provided we assume that parameters in each of the stages are a priori independent then simple tower rules can be used to calculate the mean and variance of a particular effect as a function of the posterior means and variances of the probabilities in the florets [68]. It follows that even if prior-to-posterior inferences on the stage probability vectors are more refined—perhaps estimated numerically—then we can always simply calculate excellent approximations of the moments of these probabilities—for example by sampling—and calculate the mean and variance of quantities we want. We will do this using a real system below.

Return to the CHDS dataset we analysed throughout Chapter 7 and assume CEG B obtained through the AHC algorithm and depicted in Figure 7.5a—here repeated as Figure 8.4—represents the correct choice of explanatory model. Now let us make the bold assumption that this CEG is causal as in Definition 8.4 with respect to the manipulation we consider below. We can thus demonstrate how the causal inference techniques above might be employed to evaluate the promise of various ways in which we might choose to intervene in the system to try to improve the conditions of these children. More detailed discussions of this example are given in [23].

In the CEG B, children from a family of high social background and a low number of life events are in position w_6 and hence have a different probability of hospital admission X_{h} to the other individuals. Children from socially advantaged families with an average number of life events are in the same position w_7 as children from socially disadvantaged families with a high economic background and a low or average number of life events, as are children from a low economic background with a low number of life events. All individuals with a high number of life events are in the same position w_8 irrespective of their social or economic background, and are in the same position as an individual from a low social and economic background with only an average number of life events.

Reading the CEG B above as potentially causal, we see that the economic situation X_{e} of a family seems to have no effect on the number of life events X_{l}

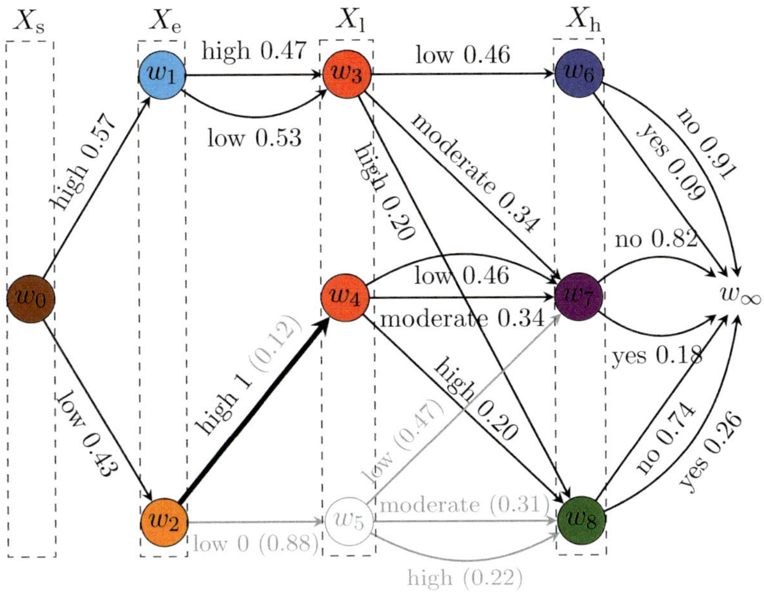

FIGURE 8.4
The CEG B from Figure 7.5a representing the CHDS, now subject to an atomic manipulation.

for families from a higher social background. However, in a family from a lower social background the economic situation seems to provoke the number of life events to increase. So we might conjecture that were we to intervene to improve the economic situation of a poor family then this would help prevent life events the monitored child experienced. More explicitly, improving the disadvantaged economic situation of a child's family from *low* to $\text{do}(X_\text{e} = \text{high})$ would give rise to the same distribution of life events as it would had the child been in one of the three other categories where at least one of the variables X_e or X_s takes the value *high*.

Using the semantics introduced in Section 8.2, we can now calculate the effect of this intervention. Let thus $\hat{e} = (w_2, w_4)$ denote the edge we intervene on—the thick depicted edge in Figure 8.4—and let p be the probability mass function of the idle CEG B before intervention. The probabilities $p(\omega)$ of every atomic event ω represented by one of the 24 root-to-sink paths in the CEG can again be calculated multiplying the attached edge probabilities. We repeat the estimated mean values of these in Figure 8.4 as inferred from Table 8.1. Our variable of interest is now the event 'admission to hospital for children from a low social background'. So under this probability mass function, the probability of this event in the idle system can simply be calculated to be $P(X_\text{h} = \text{yes} \mid X_\text{s} = \text{low}) = 0.239$ where P denotes the probability measure associated to p. Using the reported Dirichlet distributions in Table 8.1, the variance of this estimate is $2.7 \cdot 10^{-4}$.

The new manipulated CEG $\mathrm{B}_{-\hat{e}}$ where the edge $\hat{e}$ has been cancelled is depicted in Figure 8.4 as part of the idle CEG B. Those edges that are erased after intervention have been greyed out. These are precisely all edges entering and leaving the position w_5 whose history has been associated with low social background. Now the total effect of the intervention on the edge $\hat{e}$ is given by the new probability mass function $p_{-\hat{e}}$ on the remaining 18 root-to-leaf paths. Because now all children from socially disadvantaged families are given economic support, the unfolding given by $\hat{e}$ is the only possible development from position w_2 and has been assigned probability one. For children from these families, we can thus redo the calculations above, using again the mean values of edge probabilities which have not been greyed out. We hence obtain the new probability of being admitted to hospital to be $P_{-\hat{e}}(X_{\mathrm{h}} = \text{yes} \mid X_{\mathrm{s}} = \text{low}) = 0.194$ after intervention, where $P_{-\hat{e}}$ denotes the probability measure associated to $p_{-\hat{e}}$. As a consequence, this intervention has significantly reduced the probability of hospital admission for children from a low social background! The variance of this new probability can hence be calculated to stay roughly the same at $3.3 \cdot 10^{-4}$.

We can now use this result as a starting point to suggest to domain experts to design a properly randomised trial-intervention in the CHDS population which gives economic support to a group of children from a low social background. We then collect new data about these children and a control group which has not been given support. Using the manipulated CEG $\mathrm{B}_{-\hat{e}}$ and its associated probability distributions as a prior, we can thus again use one of the model selection techniques from Chapter 6 to see whether or not our manipulation of the system has had the desired effect.

Of course the tree-atomic manipulation we considered above might not be the only feasible one. For instance, if it were possible to intervene on children from a high social background by ensuring a low number of life events this would ensure the child reached the position w_6 in that representation which has the lowest probability of hospital admission. This would support the possible health benefits from pursuing a policy which enacted this.

The causal interpretation of the CEG C depicted Figure 7.5b in the previous chapter is broadly similar to that just discussed for CEG B in Figure 8.4. Again the economic situation here seems to have no effect on the number of life events X_{l} for families from a higher social background. But in a family from a lower social background the economic situation does seem to affect the number of life events that occur. Hence we again infer, in interpreting the CEG C causally, that improving the economic situation from low to high for such children would give rise to the same distribution of life events as it would had the child been in one of the three other categories. Both CEGs B and C suggest that children suffering a high number of life events are in the same

TABLE 8.1
Prior and posterior probability distributions and data associated with each stage of the CEG depicted in Figure 8.4. The greyed out values for w_5 correspond to those which will be manipulated by the causal intervention. Compare also Table 7.5.

Stage	Prior	$\boldsymbol{y}_i$, $i = 0, \ldots, 8$	Posterior
w_0	$\text{Dir}(1.5, 1.5)$	$(507, 383)$	$\text{Dir}(508.5, 384.5)$
w_1	$\text{Dir}(0.75, 0.75)$	$(237, 270)$	$\text{Dir}(237.75, 270.75)$
w_2	$\text{Dir}(0.75, 0.75)$	$(46, 337)$	$\text{Dir}(46.75, 337.75)$
w_3, w_4	$\text{Dir}(0.75, 0.75, 0.75)$	$(255, 190, 108)$	$\text{Dir}(255.75, 190.75, 108.75)$
w_5	$\text{Dir}(0.25, 0.25, 0.25)$	$(74, 105, 158)$	$\text{Dir}(74.25, 105.25, 158.25)$
w_6	$\text{Dir}(0.25, 0.25)$	$(213, 21)$	$\text{Dir}(213.25, 21.25)$
w_7	$\text{Dir}(0.625, 0.625)$	$(235, 50)$	$\text{Dir}(235.625, 50.625)$
w_8	$\text{Dir}(0.625, 0.625)$	$(273, 98)$	$\text{Dir}(273.625, 98.625)$

position regarding hospital admission regardless of their social or economic background.

However, the causal inferences we might make from the best scoring CEG G in the previous chapter, depicted in Figure 7.10a and obtained from the DP algorithm, are rather different. Here, the order of the hospital admissions X_h and the life event X_l have been reversed so that life events are expressed as a consequence of a child's health. This would imply that for children in position w_5 in that representation—that is, children who had a high social background but were poor, or had low social background but were not poor—supporting such families so that the health of the child was improved, so that she no longer needed to be admitted to hospital, would help prevent life events (like divorce) happening. This also is a plausible hypothesis since having a sick child will put stress on the family and may lead to an increased number of life events. Notice however that the causal predictions associated with these two models are quite different. One encourages giving support to families to avoid life events whilst the other encourages support for families in addressing the health needs of their child.

Clearly, the assertion that any of these CEGs is causal is a speculative one. We will discuss some initial methods on how to *discover* whether or not such a model is causal in the next section. Furthermore, we reported the Bayes Factor between the two MAP CEGs to be only 1.14, so the observational data does not distinguish between the two models very strongly. In this case since the two models have almost the same posterior probabilities, on the basis of these data we certainly cannot come to the any strong conclusion about a causal directionality. However were the BF ratio much larger then it might be possible to make stronger deductions.

Because of the speculative nature of the causal embellishment above, typically such causal hypotheses are most effectively used to prompt the further investigation of the process through interventional experiments which can more reliably distinguish between them. Thus to find support of the CEG B above we might propose a trial intervention on a small sub-population of the vulnerable families designed to mitigate life events and then observe the impact within this sub-population on subsequent hospital admission of the child. Were either of the models given by CEG C or CEG G causal, then such an intervention would be predicted to have no effect. Similarly, an intervention to reduce child hospital admissions when possible (through for example home treatments) on the vulnerable population should under the CEG G help mitigate the number of life events in the family, whilst under either the model represented by CEG B or CEG C such interventions would be predicted to have no effect.

More subtly, the fact that our exploratory technique has led us to discover two different competing causal hypotheses (CEG C and CEG G) that seem to be well supported might also encourage us to entertain models that lie outside the original class: see [23] for further discussion.

8.3.3 Backdoor theorems

Often when we observe data informing a causal system we are only able to observe certain marginal probabilities. This can make the estimation of a causal effect rather more difficult than described in the previous section. In fact, sometimes even as a sample size tends to infinity we might not be able to precisely estimate what the effect of a particular intervention on a given variable of interest might be. From the formulae given in Section 8.1 we can see that for a BN causal effects can typically only be calculated if we can write the causal effect as a function of the parameters of the unobserved system. We call such effects *identifiable*. And even then these estimates may not be consistent [86]. We outline below how we can inherit this problem for CEGs, too.

In a partially observed system, we typically infer a certain random vector measurable with respect to a non-trivial partition $A_1, A_2, \ldots, A_m$ of the atoms Ω. For instance, we might have observed data $\boldsymbol{Y} = \boldsymbol{y}$ where the vector $\boldsymbol{y}$ is of length strictly less than the number $m > r$ of atoms in the space of interest, and A_i corresponds to the event $\boldsymbol{Y}_i = \boldsymbol{y}_i$ for $i = 1, \ldots, r$. Then we would need to estimate the probabilities

$$\psi_i = P(A_i) = \sum_{\omega \in A_i} p(\omega) \tag{8.11}$$

for all $i = 1, \ldots, r$, where each atomic probability $p(\omega)$ in this formula is a monomial in the stage floret probabilities. By the results of Section 5.1, the expression in (8.11) is thus a sum of products of Dirichlet distributions. If we randomly sample over this partition, we then obtain a multinomial sample over the vector of probabilities $\boldsymbol{\psi} = (\psi_1, \psi_2, \ldots, \psi_r)$. In particular, we will be able to specify a posterior probability distribution to this probability vector to an arbitrary degree of precision given we sample enough units. However, the dimension $r-1$ of $\boldsymbol{\psi}$ can often be much smaller than the number of parameters we need to estimate, even with a large amount of stage structure. So even with a large dataset we might not be able to estimate all parameters in a model on Ω. Furthermore, each component ψ_i is now a polynomial function of the floret probabilities. But the floret probability vectors will be the arguments of the probabilities needed to evaluate the predictive distribution of the total effect we need. This can make inversion—and hence the calculation of a causal effect—challenging or impossible to perform. A detailed discussion of this issue for BNs which carries over to CEGs is given in [77].

In [104, 105] we have analysed several settings where a given idle CEG was sufficient for us to determine when such inversions were possible. The methodologies involved in this type of inference are often labelled *backdoor theorems* for CEGs—in analogy to the terminology used in Bayesian networks [77]. Because these results are rather technical we will not cover any of the details here.

An alternative method to determine whether the effect of a specific causal manipulation is identifiable is to use techniques from computer algebra, partiticularly substitution theory [105]. For simple problems, this alternative method seems to work well. One observation we can make is that it appears to be only slightly more difficult to determine whether a cause is identifiable within the very wide class of CEG causal models than it is to determine this for causal models framed within the much more restrictive class of BNs.

8.4 Causal discovery algorithms for CEGs

The established inferential methods we discussed above of course pertain only to situations where certain features of the distribution of the observed system—explicitly, various factorisations of marginal and conditional densities over subspaces of the idle space, often expressed as collections of conditional independence statements over measurable variables within the space—are already known, or at least plausibly conjectured. Of course this is often not the case. Then causal inference needs to take place in two phases. The first phase, a preprocessing phase, searches over a class of models in an observed system to find within the data evidence for collections of factorisations of densities that might embody plausible causal hypotheses. Only then can the mathematical apparatus described above be applied to construct hypotheses of what might happen under certain controls. This two-phase approach to inference, known as *causal discovery*, has been intrinsic to early causal research and has in fact motivated much of its development. It has now successfully been employed across a wide range of domains, for instance through the PC algorithm [100].

However, causal discovery algorithms are not currently available to search many important and more general classes of causal models, especially those associated with collections of hypotheses about factorisations of probability mass functions over directed probability trees, or CEGs. So the final plank in developing causal technology for staged trees in analogy to BNs is to provide fast causal discovery algorithms. This programme of work was initially frustrated because until very recently the necessary characterisation of statistical equivalence classes of staged trees and CEGs was not available. However, by the use of polynomial algebraic methods [47] these equivalence classes have now been classified: see Section 4.2. This was a vital component of this research since causal deductions from an observed idle system can only be legitimate if these deductions are consistent with an equivalence class of models stating equivalent data generating processes in the observed system.

With the causal algebra for CEGs discussed in the previous section in place, we are now ready to investigate when these formulae might actually apply. Suppose thus that instead of using a CEG to use *elicited* domain knowledge to extend the graph or tree so that it also represents the probability distribution of what might happen under a set of selected controls, we instead try to use a CEG we found through model selection to *discover* various extensions and selected controls that are promising candidates on which the causal algebraic extensions—as in Definition 8.4—might indeed hold. Might this be possible?

Again the causal discovery programmes seeded by the work in BNs [77, 100] have reflected on this issue and many authors have come up with various fascinating ways of addressing this question. Despite this type of inference being by its very nature speculative and exploratory, this has not diminished its successful use in a wide range of applications in physical science, social science, medicine and economics [10]. Causal inference here follows the typical scientific protocol of searching for symmetries in the data to find interesting conjectures about how features of a given problem might interact with one another. Subsequently scientists are then encouraged and emboldened to construct controlled experiments to formally test out whether these driving mechanisms indeed exist. Within the statistical paradigm, this activity therefore is often labelled as *exploratory data analysis* [52]—here specifically concerned with the construction of plausible causal conjectures about how a data stream is being created.

One of the central ideas in graph-based causal discovery algorithms is to conjecture that edge directions that appear in a fitted BN suggest that a vertex X at the beginning of a directed edge $X \rightarrow Y$ is a potential putative 'cause' of the receiving 'effect' vertex Y of that edge. This might then prompt a scientist to design an experiment where she really can manipulate that causal variable to various of its levels—enforcing the control $\text{do}(X = x)$ for a collection of states $x \in \mathbb{X}$—and see whether the effect really was what it was predicted to be under hypotheses about effects like (8.2). Compare also the development in Section 8.3 above.

Considerable caution needs to be exerted in making causal deductions in this way. We discussed earlier that two BNs with the same vertices and sharing different edge connections could be statistically equivalent to each other, so could represent the same collection of modelling assumptions. Recall from Section 2.3 that this means that the family of probability distributions they represent are identical. Two statistically equivalent models will always give identical sample distributions and so will 'explain' the data equally well whatever that data turned out to be. So it would be inappropriate to deduce from a BN represented by a DAG $\mathcal{G}$ that a directionality $X \rightarrow Y$ was causal if there was another statistically equivalent DAG $\mathcal{G}'$ with that edge $X \leftarrow Y$ going in the opposite direction!

Proceeding with this argument it would follow that a *necessary* condition for such a causal relationship to exist is that *all* edges in the equivalence class of the discovered BN were in the same direction: see also our discussion in Section 8.1 about causes happening before an effect. It was proved that these were exactly those edges in its *essential* graph [77]. The essential graph of a DAG is equal to its pattern where additionally all edges are directed which have always the same directionality across the whole statistical equivalence class: compare Section 2.3. For example for a decomposable BN, there are no such directed edges. Pearl called these relationships *direct potential causes.* Typically these require that an effect node Y has come after a collider $U \rightarrow X \leftarrow V$, so after two disconnected vertices sharing the same child. These potential causes would then correspond to the vertices along a directed path $X \rightarrow Y$ in the essential graph into that effect variable—and a direct potential cause would be the vertex X directly before the effect on that path.

Although this necessary condition has been sufficient for many authors, Pearl argued—in our view compellingly—that even this was not enough. His view is that what was observed was simply a margin of a much larger BN with many unobserved variables. If X was to be designated a cause of Y in the original BN it must surely be a designated cause of Y in any such larger more descriptive model. However this effectively removes a causal designation to any cause of an effect which is only a potential cause because it happens to be one of a collider pair. For instance, if the directionality $X \rightarrow Y$ was inferred from an essential graph then as a margin of a larger system, this edge might still be replaced by $X \leftarrow Z \rightarrow Y$ for some formerly unobserved random variable Z—but this new DAG is decomposable and so we cannot infer an unambiguous directionality from X to Y in the larger system.

How can these ideas be transferred to CEGs? If this is to happen then we first need to be able to determine when a causal situation appears before the set of situations which constitute the set of events associated with its effects. By analogy to the argument above for BNs, for this situation to be labelled as a potential cause this would need to be true for all statistically equivalent CEGs: so whenever an event associated with a vertex v is downstream of the event associated with another vertex w then in all statistically equivalent representations of this model, the vertex depicting the event associated with v must be downstream of the event associated with w.

Importantly, the characterisation of equivalence classes of staged trees as presented in Section 4.2 would not have been possible if it had not been for notions translated from polynomial algebra and algebraic geometry to tree graphs. Through this, an analysis of the interpolating polynomial of a staged tree model is central to the understanding of possible applications of the swap and resize operators, which in turn are key to the analysis of possible orderings of events within a model. As a consequence, the ordering of up- and downstream events in a CEG—which is analogous to that induced by collider structures ordering problem variables in the essential graph—here arises from algebraic rather than graphical properties of the model. Because, by the development in Section 3.3, discrete BNs form a subclass of the model class of staged trees, it is relatively straightforward to develop concepts which are analogous to those of *genuine* and *putative* cause in DAGs [77]. Building on our characterisation of statistically equivalent staged trees, we are thus at last in a position to study in ongoing research *causally equivalent* staged trees, that is trees which suggest identical causal hypotheses. We can therefore discover the potential causes of an effect simply by discovering the best fitting CEG and then checking out its equivalence classes. Because the CEG is more nuanced than a DAG there are often many more nuanced ways for this directionality to be unambiguous. We illustrate the ideas above with the revisiting of the CHDS model from Chapter 7 in [43, 44]. For instance, we find that there is a high scoring SCEG which across all of its statistically equivalent representations always orders the variable hospital admission before life events. Compare also the discussion in Section 8.3.2 above.

At the time of writing of this book this research programme has not yet been completed. However, very promising early analyses indicate that an entirely general theory of causal discovery for CEGs is possible. We plan to report some of this work in the coming year.

8.5 Exercises

Exercise 8.1. *For the manipulated degenerate probability tree $\mathcal{T}_{\hat{e}}$ in Figure 8.1, draw the non-degenerate probability tree $\mathcal{T}_{-\hat{e}}$ which has the same associated probability mass function, $p_{\hat{e}} = p_{-\hat{e}}$.*

Exercise 8.2. *Repeat Exercise 8.1 for the staged tree in Figure 8.2. In addition, also draw the corresponding idle CEG and the manipulated degenerate and non-degenerate CEGs.*

Exercise 8.3. *Consider again the accommodation example from Section 8.2.2. Calculate the probability of a student leaving the city of Coventry, assuming she was initially forced to live there, in two settings:*

1. *using (8.2) and the DAG in Figure 8.3, and*
2. *using (8.6) and the staged tree in Figure 8.2.*

Hint: a solution is provided in [46].

Exercise 8.4. *Investigate in Figure 8.2 which situations a policy maker might be able to manipulate and which edges can be logically enforced in the context represented by the idle tree. Hence determine the set A from Definition 8.4 for which this staged tree is A-causal.*

Exercise 8.5. *In the CHDS example from Section 8.3.2, can you think of atomic manipulations—for instance of the variable life events—for which the causal formula (8.2) might hold or not hold, respectively?*

References

[1] Aggarwal, C. C. and C. K. Reddy (Eds.) (2014). *Data Clustering: Algorithms and Applications.* CRC Press.

[2] Altomare, D., G. Consonni, and L. La Rocca (2013). Objective Bayesian Search of Gaussian Directed Acyclic Graphical Models for Ordered Variables with Non-Local Priors. *Biometrics 69*(2), 478–487.

[3] Barbu, V. S. and N. Limnios (2008). *Semi-Markov Chains And Hidden Semi-Markov Models Toward Applications: Their Use In Reliability And DNA Analysis*, Volume 191. Springer.

[4] Barclay, L. M., R. A. Collazo, J. Q. Smith, P. Thwaites, and A. Nicholson (2015). The dynamic Chain Event Graph. *Electronic Journal of Statistics 9*(2), 2130–2169.

[5] Barclay, L. M., J. L. Hutton, and J. Q. Smith (2013). Refining a Bayesian Network using a Chain Event Graph. *International Journal of Approximate Reasoning 54*(9), 1300–1309.

[6] Barclay, L. M., J. L. Hutton, and J. Q. Smith (2014). Chain Event Graphs for informed missingness. *Bayesian Analysis 9*(1), 53–76.

[7] Barons, M. J., S. K. Wright, and J. Q. Smith (2017). Eliciting probabilistic judgements for integrating decision support systems. In J. Quigley, A. Morton, and L. Dias (Eds.), *Elicitation of Preferences and Uncertainty: Processes and Procedures.* Springer.

[8] Berger, J. O. and L. R. Pericchi (2001). Objective Bayesian Methods for Model Selection: Introduction and Comparison. In P. Lahiri (Ed.), *Model selection*, Volume 38 of *Lecture Notes–Monograph Series*, pp. 135–207. Beachwood, OH: Institute of Mathematical Statistics.

[9] Bernardo, J. M. and A. F. M. Smith (2004). *Bayesian Theory.* Wiley series in probability and mathematical statistics. Chichester: John Wiley.

[10] Berzuini, C., A. P. Dawid, and L. Bernardinell (2012). *Causality: Statistical Perspectives and Applications.* Wiley Series in Probability and Statistics. Wiley.

[11] Boettcher, S. G. and C. Dethlefsen. (2013). ***deal**: Learning Bayesian Networks with Mixed Variables.* R-package version 1.2-37, available at `https://CRAN.R-project.org/package=deal`.

[12] Boutilier, C., N. Friedman, M. Goldszmidt, and D. Koller (1996). Context-specific independence in bayesian networks. In E. Horvitz and F. Jensen (Eds.), *12th Conference on Uncertainty in Artificial Intelligence (UAI 96)*, Uncertainty in Artificial Intelligence, San Francisco, pp. 115–123. Morgan Kaufmann Publishers Inc.

[13] Call, H. J. and W. A. Miller (1990). A comparison of approaches and implementations for automating decision analysis. *Reliability Engineering and System Safety 30*, 115–162.

[14] CHDS (2014). Christchurch Health and Development Study. Online resource available at `http://www.otago.ac.nz/christchurch/research/healthdevelopment/` accessed February 8, 2017.

[15] Chipman, H., E. I. George, and R. E. McCulloch (2001). The practical implementation of Bayesian model selection. In P. Lahiri (Ed.), *Model selection*, Volume 38 of *Lecture Notes–Monograph Series*, pp. 65–116. Beachwood, OH: Institute of Mathematical Statistics.

[16] Collazo, R. A. (2017). *The Dynamic Chain Event Graph.* Ph. D. thesis, University of Warwick, Department of Statistics.

[17] Collazo, R. A. and J. Q. Smith (2015). A New Family of Non-Local Priors for Chain Event Graph Model Selection. *Bayesian Anal. 11*(4), 1165–1201.

[18] Collazo, R. A. and P. G. Taranti (2017). *`ceg`: Chain Event Graph.* R-package version 0.1.0, available at `https://CRAN.R-project.org/package=ceg`.

[19] Consonni, G., J. J. Forster, and L. La Rocca (2013). The whetstone and the alum block: Balanced objective Bayesian comparison of nested models for discrete data. *Statist. Sci. 28*(3), 398–423.

[20] Consonni, G. and L. La Rocca (2011). On moment priors for Bayesian model choice with applications to directed acyclic graphs. In J. M. Bernardo, M. J. Bayarri, J. O. Berger, A. P. Dawid, D. Heckerman, A. F. M. Smith, and M. West (Eds.), *Bayesian Statistics 9 – Proceedings of the Ninth Valencia International Meeting*, pp. 63–78. Oxford University Press.

[21] Cowell, R. G. and A. P. Dawid (1992). Fast retraction of evidence in a probabilistic expert system. *Statistics and Computing 2*(1), 37–40.

[22] Cowell, R. G., A. P. Dawid, S. L. Lauritzen, and D. J. Spiegelhalter (2007). *Probabilistic networks and expert systems.* Statistics for engineering and information science. New York ; London: Springer.

[23] Cowell, R. G. and J. Q. Smith (2014). Causal discovery through MAP selection of stratified chain event graphs. *Electronic Journal of Statistics 8*(1), 965–997.

[24] Cuthbertson, I. M. (2004). Prisons and the Education of Terrorists. *World Policy Journal 21*(3), 15–22.

[25] Darwiche, A. (2003). A differential approach to inference in Bayesian networks. *J. ACM 50*(3), 280–305 (electronic).

[26] Darwiche, A. (2009). *Modeling and Reasoning with Bayesian Networks.* Cambridge University Press, Cambridge.

[27] Dawid, A. P. (1999). The Trouble with Bayes Factors. Technical report, University College London.

[28] Dawid, A. P. (2011). Posterior model probabilities. In P. S. Bandyopadhyay and M. R. Forster (Eds.), *Philosophy of Statistics*, Volume 7, pp. 607–630. Amsterdam: North-Holland.

[29] Drton, M., B. Sturmfels, and S. Sullivant (2009). *Lectures on algebraic statistics*, Volume 39 of *Oberwolfach Seminars.* Birkhäuser Verlag, Basel.

[30] Fergusson, D., M. Dimond, L. Horwood, and F. Shannon (1984). The utilisation of preschool health and education services. *Social Science & Medicine 19*(11), 1173–1180.

[31] Fergusson, D., L. Horwood, A. Beautrais, and F. Shannon (1981). Health care utilisation in a New Zealand birth cohort. *Community Health Studies 5*(1), 53–60.

[32] Fergusson, D. M., L. J. Horwood, and F. T. Shannon (1986). Social and Family Factors in Childhood Hospital Admission. *Journal of Epidemiology and Community Health 40*(1), 50–58.

[33] Freeman, G. and J. Q. Smith (2011a). Bayesian MAP model selection of Chain Event Graphs. *Journal of Multivariate Analysis 102*(7), 1152–1165.

[34] Freeman, G. and J. Q. Smith (2011b). Dynamic staged trees for discrete multivariate time series: Forecasting, model selection and causal analysis. *Bayesian Analysis 6*(2), 279–305.

[35] French, S. and D. R. Insua (2010). *Statistical Decision Theory: Kendall's Library of Statistics 9.* Wiley.

[36] Garcia, L. D., M. Stillman, and B. Sturmfels (2005). Algebraic geometry of Bayesian networks. *J. Symbolic Comput. 39*(3-4), 331–355.

[37] Geiger, D. and D. Heckerman (1997). A characterization of the dirichlet distribution through global and local parameter independence. *Ann. Statist. 25*(3), 1344–1369.

[38] Geiger, D., D. Heckerman, H. King, and C. Meek (2001). Stratified exponential families: graphical models and model selection. *Ann. Statist. 29*(2), 505–529.

[39] Geiger, D., C. Meek, and B. Sturmfels (2006). On the toric algebra of graphical models. *Ann. Statist. 34*(3), 1463–1492.

[40] Geiger, D. and J. Pearl (1993). Logical and algorithmic properties of conditional independence and graphical models. *Ann. Statist. 21*(4), 2001–2021.

[41] Gibilisco, P., E. Riccomagno, M. P. Rogantin, and H. P. Wynn (Eds.) (2010). *Algebraic and Geometric Methods in Statistics.* Cambridge University Press, Cambridge.

[42] Goldstein, M. and D. Wooff (2007). *Bayes Linear Statistics, Theory and Methods.* Bayes Linear Statistics. Wiley.

[43] Görgen, C. (2017). *An algebraic characterisation of staged trees: their geometry and causal implications.* Ph. D. thesis, University of Warwick, Department of Statistics.

[44] Görgen, C., A. Bigatti, E. Riccomagno, and J. Q. Smith (2017). Discovery of statistical equivalence classes using computer algebra. Preprint available from `arXiv:1705.09457v1 [math.ST]`.

[45] Görgen, C., M. Leonelli, and J. Q. Smith (2015). A Differential Approach for Staged Trees. In *Symbolic and Quantitative Approaches to Reasoning with Uncertainty, Proceedings*, Lecture Notes in Artificial Intelligence, pp. 346–355. Springer.

[46] Görgen, C. and J. Q. Smith (2016). A differential approach to causality in staged trees. In *Proceedings of the Eighth International Conference on Probabilistic Graphical Models*, Volume 52 of *JMLR Workshop and Conference Proceedings*, pp. 207–215.

[47] Görgen, C. and J. Q. Smith (2017). Equivalence Classes of Staged Trees. *Bernoulli (forthcoming).* Preprint available from `arXiv:1512.00209v3 [math.ST]`.

[48] Hannah, G., L. Clutterbuck, and J. Rubin (2008). Radicalization or rehabilitation. Understanding the challenge of extremist and radicalized prisoners. Technical Report TR 571, RAND Corporation.

[49] Heard, N. A., C. C. Holmes, and D. A. Stephens (2006). A quantitative study of gene regulation involved in the immune response of anopheline mosquitoes: An application of bayesian hierarchical clustering of curves. *Journal of the American Statistical Association 101*(473), 18–29.

[50] Heckerman, D. (1998). In M. I. Jordan (Ed.), *Learning in Graphical Models*, Chapter A Tutorial on Learning with Bayesian Networks, pp. 301–354. Dordrecht, The Netherlands: Kluwer Academic Publishers.

[51] Heckerman, D. and D. Geiger (1995). Learning Bayesian Networks: A Unification for Discrete and Gaussian Domains. In *UAI '95: Proceedings of the Eleventh Annual Conference on Uncertainty in Artificial Intelligence, Montreal, Quebec, Canada, August 18-20*, pp. 274–284.

[52] Hoaglin, D. C., F. Mosteller, and J. W. Tukey (Eds.) (2000). *Understanding Robust and Exploratory Data Analysis*. Wiley Classics Library. Wiley-Interscience, New York. Revised and updated reprint of the 1983 original.

[53] Jaeger, M. (2004). Probabilistic decision graphs – combining verification and AI techniques for probabilistic inference. *International Journal of Uncertainty, Fuzziness and Knowledge-Based Systems 12*, 19–42.

[54] Jaeger, M., J. D. Nielsen, and T. Silander (2006). Learning probabilistic decision graphs. *International Journal of Approximate Reasoning 42*(1-2), 84–100.

[55] Jain, A. K., M. N. Murty, and P. J. Flynn (1999). Data clustering: A review. *ACM Computing Surveys 31*(3), 264–323.

[56] Jensen, F. V. and F. Jensen (1994). Optimal Junction Trees. In M. Kaufmann (Ed.), *Proceedings of the 10th Conference on Uncertainty in Artifical Intelligence*, Volume 10, San Mateo, CA.

[57] Jensen, F. V. and T. D. Nielsen (2007). *Bayesian Networks and Decision Graphs*. Bayesian Networks and Decision Graphs. Springer.

[58] Johnson, V. E. and D. Rossell (2010). On the use of non-local prior densities in bayesian hypothesis tests. *Journal of the Royal Statistical Society Series B-Statistical Methodology 72*, 143–170.

[59] Johnson, V. E. and D. Rossell (2012). Bayesian model selection in high-dimensional settings. *Journal of the American Statistical Association 107*(500), 649–660.

[60] Jordan, J. and N. Horsburgh (2006). Spain and islamist terrorism: Analysis of the threat and response 1995-2005. *Mediterranean Politics 11*(2), 209–229.

[61] Kass, R. E. and A. E. Raftery (1995). Bayes Factors. *Journal of the American Statistical Association 90*(430), 773–795.

[62] Korb, K. B. and A. E. Nicholson (2011). *Bayesian Artificial Intelligence* (2nd ed.). Chapman & Hall/CRC computer science and data analysis series. Boca Raton, FL: CRC Press.

[63] Lauritzen, S. L. (1996). *Graphical models*, Volume 17 of *Oxford Statistical Science Series*. The Clarendon Press, Oxford University Press, New York. Oxford Science Publications.

[64] Lauritzen, S. L. (2001). Causal inference from graphical models. In *Complex Stochastic Systems (Eindhoven, 1999)*, Volume 87 of *Monogr. Statist. Appl. Probab.*, pp. 63–107. Chapman & Hall/CRC, Boca Raton, FL.

[65] Lauritzen, S. L., A. P. Dawid, B. N. Larsen, and H.-G. Leimer (1990). Independence properties of directed Markov fields. *Networks 20*(5), 491–505. Special issue on influence diagrams.

[66] Lauritzen, S. L. and T. S. Richardson (2002). Chain graph models and their causal interpretations. *J. R. Stat. Soc. Ser. B Stat. Methodol. 64*(3), 321–361.

[67] Leonelli, M., C. Görgen, and J. Q. Smith (2017). Sensitivity analysis in multilinear probabilistic models. *Information Sciences 411*, 84–97.

[68] Leonelli, M. and J. Q. Smith (2015). Bayesian decision support for complex systems with many distributed experts. *Annals of Operations Research 235*(1), 517–542.

[69] Liverani, S., J. Cussens, and J. Q. Smith (2010). Searching a multivariate partition space using max-sat. In F. Masulli, L. Peterson, and R. Tagliaferri (Eds.), *Computational Intelligence Methods for Bioinformatics and Biostatistics*, Volume 6160 of *Lecture Notes in Computer Science*, pp. 240–253. Springer Berlin Heidelberg.

[70] Medhi, J. (1994). *Stochastic Processes*. New Age International.

[71] Mond, D., J. Smith, and D. van Straten (2003). Stochastic factorizations, sandwiched simplices and the topology of the space of explanations. *R. Soc. Lond. Proc. Ser. A Math. Phys. Eng. Sci. 459*(2039), 2821–2845.

[72] Neapolitan, R. E. (2004). *Learning Bayesian Networks*. Harlow: Prentice Hall.

[73] Neumann, P. E. (2010, July). Prisons and terrorism: Radicalisation and de-radicalisation in 15 countries. Technical report, International Centre for the Study of Radicalisation and Political Violence, London.

[74] O'Hagan, A., C. Buck, A. Daneshkhah, J. Eiser, P. Garthwaite, D. Jenkinson, J. Oakley, and T. Rakow (2006). *Uncertain Judgements: Eliciting Experts' Probabilities*. Statistics in Practice. Wiley.

[75] Pachter, L. and B. Sturmfels (2005). *Algebraic Statistics for Computational Biology*. Cambridge University Press, New York.

[76] Pearl, J. (1988). *Probabilistic Reasoning in Intelligent Systems: Networks of Plausible Inference.* San Francisco, CA, USA: Morgan Kaufmann Publishers Inc.

[77] Pearl, J. (2000). *Causality* (First ed.). Cambridge University Press, Cambridge. Models, reasoning, and inference.

[78] Pericchi, L. R. (2005). Model selection and hypothesis testing based on objective probabilities and bayes factors. In D. K. Rao and C. Dey (Eds.), *Handbook of Statistics*, Volume 25, pp. 115–149. Elsevier.

[79] Pistone, G., E. Riccomagno, and H. P. Wynn (2001a). *Algebraic Statistics*, Volume 89 of *Monographs on Statistics and Applied Probability.* Chapman & Hall/CRC, Boca Raton, FL. Computational commutative algebra in statistics.

[80] Pistone, G., E. Riccomagno, and H. P. Wynn (2001b). Gröbner bases and factorisation in discrete probability and Bayes. *Stat. Comput. 11*(1), 37–46.

[81] Puch, R. O. and J. Q. Smith (2002). Finds: A training package to assess forensic fibre evidence. *Advances in Artificial Intelligence*, 420–429.

[82] R Core Team (2016). `R`: A Language and Environment for Statistical Computing. Available at `https://www.R-project.org`.

[83] Rao, C. R. (1995). A review of canonical coordinates and an alternative to correspondence analysis using Hellinger distance. *Questiio 19*(1-3), 23–63.

[84] Rigat, F. and J. Q. Smith (2012). Isoseparation and Robustness in Finite Parameter Bayesian Inference. *Annals of the Institute of Statistical Mathematics 64*(3), 495–519.

[85] Robins, J. (1986). A new approach to causal inference in mortality studies with a sustained exposure period—application to control of the healthy worker survivor effect. *Math. Modelling 7*(9-12), 1393–1512. Mathematical models in medicine: diseases and epidemics, Part 2.

[86] Robins, J. M., R. Scheines, P. Spirtes, and L. Wasserman (2003). Uniform consistency in causal inference. *Biometrika 90*(3), 491–515.

[87] Rubin, D. B. (1978). Bayesian inference for causal effects: the role of randomization. *Ann. Statist. 6*(1), 34–58.

[88] Salmerón, A., A. Cano, and S. Moral (2000). Importance sampling in Bayesian networks using probability trees. *Comput. Statist. Data Anal. 34*(4), 387–413.

[89] Schachter, R. D. (1988). Probabilistic Inference and Influence Diagrams. *Operations Research 36*(4), 589–605.

[90] Shafer, G. (1996). *The Art of Causal Conjecture.* Artificial Management. MIT Press, Cambridge.

[91] Silander, T., P. Kontkanen, and P. Myllymaki (2007). On sensitivity of the MAP bayesian network structure to the equivalent sample size parameter. In *Proceedings of the Conference on Uncertainty in Artificial Intelligence*, pp. 360–367.

[92] Silander, T. and T.-Y. Leong (2013). A dynamic programming algorithm for learning chain event graphs. In J. Fürnkranz, E. Hüllermeier, and T. Higuchi (Eds.), *Discovery Science*, Volume 8140 of *Lecture Notes in Computer Science*, pp. 201–216. Springer Berlin Heidelberg.

[93] Silander, T. and P. Myllymaki (2006). A simple approach for finding the globally optimal bayesian network structure. In *Proceedings of the Twenty-Second Conference Annual Conference on Uncertainty in Artificial Intelligence (UAI-06)*, Arlington, Virginia, pp. 445–452. AUAI Press.

[94] Silke, A. (2011). *The Psychology of Counter-Terrorism.* Cass series on political violence. Abingdon, Oxon, England; New York: Routledge.

[95] Sivia, D. S. (1996). *Data analysis: A Bayesian tutorial.* Oxford Science Publications. The Clarendon Press, Oxford University Press, New York.

[96] Smith, J. Q. (1981). The multiparameter steady model. *Journal of the Royal Statistical Society. Series B (Methodological) 43*(2), 256–260.

[97] Smith, J. Q. (2010). *Bayesian Decision Analysis: Principles and Practice.* Cambridge; New York: Cambridge University Press.

[98] Smith, J. Q. and P. E. Anderson (2008). Conditional independence and chain event graphs. *Artificial Intelligence 172*(1), 42–68.

[99] Smith, J. Q., M. J. Barons, and M. Leonelli (2015). Decision focused inference on networked probabilistic systems: with applications to food security. In *JSM Proceedings – Section on Bayesian Statistical Science*, pp. 3220–3233.

[100] Spirtes, P., C. Glymour, and R. Scheines (1993). *Causation, Prediction, and Search* (1st ed.). MIT Press.

[101] Spivey, M. Z. (2008). A generalized recurrence for Bell numbers. *Journal of Integer Sequences 11*(2).

[102] Studený, M. (2005). *Probabilistic Conditional Independence Structures.* Information Science and Statistics. Springer, London.

[103] Taroni, F., A. Biedermann, S. Bozza, P. Garbolino, and C. Aitken (2014). *Bayesian networks for probabilistic inference and decision analysis in forensic science* (Second ed.). Statistics in Practice. John Wiley & Sons, Ltd., Chichester. With a foreword by Ian Evett.

[104] Thwaites, P. (2013). Causal identifiability via chain event graphs. *Artificial Intelligence 195*, 291–315.

[105] Thwaites, P., J. Q. Smith, and E. Riccomagno (2010). Causal analysis with Chain Event Graphs. *Artificial Intelligence 174* (12-13), 889–909.

[106] Thwaites, P. A. and J. Q. Smith (2015a). A Separation Theorem for Chain Event Graphs. Preprint available from `arXiv:1501.05215v1 [stat.ME]`.

[107] Thwaites, P. A. and J. Q. Smith (2015b). A new method for tackling asymmetric decision problems. In *Proceedings of the 10th Workshop on Uncertainty Processing (WUPES'15)*, pp. 179–190.

[108] Thwaites, P. A., J. Q. Smith, and R. G. Cowell (2008). Propagation using Chain Event Graphs. In *Proceedings of the Twenty-Fourth Conference Annual Conference on Uncertainty in Artificial Intelligence (UAI-08)*, Corvallis, Oregon, pp. 546–553. AUAI Press.

[109] Verma, T. and J. Pearl (1990). Causal networks: semantics and expressiveness. In *Uncertainty in Artificial Intelligence, 4*, Volume 9 of Mach. Intelligence Pattern Recogn., pp. 69–76. North-Holland, Amsterdam.

[110] West, M. and J. Harrison (1999). *Bayesian Forecasting and Dynamic Models* (2nd ed.). Springer series in statistics. New York: Springer.

[111] Xu, R. and D. C. Wunsch (2009). *Clustering*. IEEE series on computational intelligence. Wiley.

[112] Zwiernik, P. (2016). *Semialgebraic statistics and latent tree models*, Volume 146 of *Monographs on Statistics and Applied Probability*. Chapman & Hall/CRC, Boca Raton, FL.

Index

P

R

S

T

For Product Safety Concerns and Information please contact our
EU representative GPSR@taylorandfrancis.com Taylor & Francis
Verlag GmbH, Kaufingerstraße 24, 80331 München, Germany